Applications of high performance liquid chromatography

Applications of high performance liquid chromatography

A. Pryde
Dr. R. Maag A. G., Dielsdorf, Switzerland

and
M. T. Gilbert
Department of Chemistry, University of Edinburgh

London and New York

CHAPMAN AND HALL

First published 1979
by Chapman and Hall Ltd.,
11 New Fetter Lane, London EC4P 4EE.

© 1979 A. Pryde and M. T. Gilbert

Reprinted 1980

Published in the USA
by Chapman and Hall
in association with Methuen, Inc.
733 Third Avenue, New York, NY 10017

Typeset by Preface Ltd., Salisbury
and printed in Great Britain at the
University Press, Cambridge

ISBN 0 412 14220 1

Contents

PART II   THE APPLICATION OF HPLC IN
          PHARMACEUTICAL ANALYSIS

PART III   APPLICATION OF HPLC IN BIOCHEMICAL
           ANALYSIS

PART IV   ENVIRONMENTAL ANALYSIS BY HPLC

PART V MISCELLANEOUS APPLICATIONS

Preface

The rapidly increasing number of publications which are appearing on HPLC at this time makes it opportune to review progress in the applications field. In the first part of the book we have confined our discussion of chromatographic theory to those areas which have a direct bearing on practical results; HPLC equipment and the various modes of chromatography are discussed at more length and where possible we have tried to offer tips for avoiding some of the pitfalls of HPLC. In the succeeding sections, HPLC applications are divided into broad areas of interest (Pharmaceutical, Biochemical and Environmental analysis) and, within each section, treated according to compound class. Clearly, some of the material could have been classed in more than one section (this is especially true of some of the Pharmaceutical and Biochemical sub-sections) and we trust we have arrived at a satisfactory arrangement of the material

We have tried to cover the literature up to the Spring of 1977. Completely comprehensive coverage of HPLC applications is becoming an increasingly difficult task since, as the technique has gained rapidly in popularity over the last few years, more and more applications are to be found in publications of a more general nature. Hopefully, we have covered each section in sufficient depth to provide an idea of the most promising HPLC systems in that area. In discussing individual applications, we have usually tried to give sufficient details of the chromatographic conditions used (e.g. the column dimensions, packing material and eluent, as well as the detection system and detection limits where appropriate) to convey an idea of the potential of the method.

In summary, we hope to have succeeded in our aim of providing a straightforward and practically-oriented text on HPLC applications which will help those who actually have to solve problems by HPLC to find at least one approach suitable to their problem.

November, 1977.

Andrew Pryde,
Dielsdorf, Switzerland.

Mary T. Gilbert,
Edinburgh, Scotland.

Acknowledgments

We are especially indebted to Professor John H. Knox (Director of the Wolfson Liquid Chromatography Unit, Edinburgh University) and Dr R. A. Wall (Chemistry Department, Edinburgh University) for reading large sections of the unfinished text and for many helpful suggestions.

We are also grateful to our colleagues listed below for their many constructive comments on sections of the text within their own areas of interest: R. Aylott, Inveresk Research International, Musselburgh; P. Bristow, I.C.I. Pharmaceuticals, Macclesfield; R. Harrison, Eli Lilley Ltd., Windlesham; D. Howie, Pentland Scotch Whisky Distillers Ltd., Edinburgh; O. Jones, Bangor General Hospital, Bangor; J. Millar, Pharmaceutical Society of Great Britain, Edinburgh; P. A. Raven, Allen and Hanbury Ltd., Ware; A.G. Smith, Medical Research Council Research Centre, Carshalton; F. P. A. Vonder Mühll, Dr R. Maag, A. G., Switzerland; J. Wells, Poultry Research Centre, Edinburgh.

Finally, our thanks are in no small measure due to our secretary, Mrs Marjory Duncan.

Abbreviations and symbols

Throughout this book the pressure conversion factors used are:

$$14.7 \text{ psi} = 1 \text{ bar} = 10^5 \text{ Nm}^{-2} = 10^5 \text{ Pa} = 10^6 \text{ dyne cm}^{-2} = 0.987 \text{ std atm.}$$

Abbreviations

AUFS	absorbance units full scale
BOP	β,β'-oxydipropionitrile
DNP	dinitrophenylhydrazine
EDTA	ethylenediaminetetraacetic acid
GLC	gas-liquid chromatography
H.E.T.P.	height equivalent to a theoretical plate
HPLC	high performance liquid chromatography
i.d.	internal diameter
IR	infrared
MS	mass spectrometry
NMR	nuclear magnetic resonance
o.d.	outside diameter
ODS	octadecylsilyl
PEG	polyethylene glycol
RI	refractive index
TBA	tetrabutylammonium
TLC	thin layer chromatography
Tris	tris(hydroxymethyl) aminomethane
UV	ultraviolet

Symbols

A,B,C	constants in Knox equation
α	selectivity
D_m	diffusion coefficient
d_c	column diameter
d_p	particle diameter
ϵ	molar extinction coefficient
ϵ^0	solvent strength parameter
η	viscosity
H	plate height

h	reduced plate height
K	permeability
k	column permeability
k'	column capacity ratio
L	column length
λ	wavelength
M_{eluent}	eluent molecular weight
N	number of theoretical plates
ΔP	pressure drop down column
R_f	in TLC, the ratio of the distance travelled by the solute to that travelled by the solvent front
R_s	resolution
σ	standard deviation
T	temperature
t_0	retention time of unretained (solvent) peak
t_r	retention time of retained peak
t_1, t_2	retention distances for two peaks
u	eluent linear velocity
ν	eluent reduced velocity
V_{solute}	solute molar volume
W	base width of peak
W_{HH}	width of peak at half height
ϕ'	column resistance parameter
ψ	eluent association factor

Part I
Theory and practice of HPLC

1

Liquid chromatography: history and state of the art

The technique of high performance liquid chromatography (variously called high speed and high pressure liquid chromatography) although only around ten years old, has already made a significant contribution to pharmaceutical, biochemical, clinical and environmental analysis. The earliest example of a liquid chromatographic separation is credited to the Russian botanist Tswett who, in 1903, separated plant pigments by adsorption chromatography [1]. The technique was not followed up until the 1930s when Kuhn and Lederer [2] and Reichstein and Van Euw [3] again used adsorption chromatography for the separation of natural products. In 1941, Martin and Synge, who were subsequently awarded a Nobel Prize, described the discovery of liquid–liquid partition chromatography [4], and in the same paper laid the foundation of gas–liquid chromatography (GLC), and high performance liquid chromatography (HPLC). Martin and Synge introduced the concept of the 'height equivalent to a theoretical plate', which has since been adopted as the measure of chromatographic efficiency. They also pointed out that improved performance would be achieved in liquid chromatography by the use of smaller particles and higher pressures, and that the liquid mobile phase could be replaced by a gas. Around ten years later, in 1952, James and Martin described the first use of GLC [5] and the technique was rapidly developed during the ensuing decade. Liquid chromatography remained relatively neglected during this period although Hamilton *et al.* reported on the improvement in efficiency with reduction in particle size in the ion exchange chromatography of amino acids [6] and Snyder made a significant contribution to the understanding of adsorption chromatography [7].

In the early 1960s, Giddings showed that the theoretical framework developed for GLC applied equally well to liquid chromatography [8] and between 1967–1969, Kirkland [9], Huber [10] and Horvath, Preiss and Lipsky [11] described the first high performance liquid chromatographs. By operating at high pressure (up to 5000 psi; 34 MPa) these instruments overcame the effect of higher liquid viscosities relative to gas viscosities and gave analysis times comparable with GLC. In the early instruments, single or dual wavelength ultraviolet detectors allowed detection of nanogram amounts of suitably UV absorbing compounds, but were insensitive to compounds with little or no UV absorbance. Since then, variable wavelength UV spectrophotometers have greatly increased the scope of the UV detector; refractive index and transport detectors have been developed as more universal detectors; selective detection systems such as spectrofluorimeters and electrochemical systems are becoming common, and HPLC/mass spectrometric interfaces are now being developed.

The key to increased efficiency and improved analysis times in HPLC lay in the development of suitable support materials in which the intraparticle diffusion rates were improved by reducing the distance over which the solutes had to diffuse. Thus, Horvath et al. [11] coated impervious glass beads with a layer of ion exchange resin, and Kirkland prepared a very efficient pellicular support for liquid—liquid partition chromatography by coating glass beads with a thin layer (1—2 μm) of silica microspheres [9]. Alternatively, fully porous particles of much smaller diameter (40 μm) than those used in GLC were used as these had higher capacity than the pellicular materials. The early HPLC materials could be packed reproducibly into chromatographic columns by dry packing techniques.

Further increases in efficiency have been obtained by reducing the particle size to its present value of around 5 μm. This size represents a good compromise between efficiency, pressure drop, analysis time and reproducibility of packing. For the small particles (< 15—20 μm) dry packing techniques cannot be used and slurry packing techniques have been developed.

The range of applicability of HPLC was extended by the preparation of chemically bonded stationary phases in which the nature of the adsorbent was modified by bonding organic groups to the adsorbent surface. The first bonded phases for HPLC were prepared by Halasz et al. who reacted silica with alcohols [12] and amines [13]. Subsequently, materials with greater hydrolytic stability were prepared by bonding organosilanes to the surface of silica and a wide range of such phases is now commercially available. The chemically modified adsorbents can be used for reversed phase and ion exchange chromatography.

Compared with classical column chromatography where the columns are gravity fed and a separation can take hours or even days, HPLC can offer analysis times of 5—30 min, times which are comparable with GLC. HPLC is particularly suited to the analysis of those compounds which are not readily handled by GLC. For example, thermally labile compounds can be analyzed at ambient temperatures by HPLC, highly polar compounds can be chromatographed without prior derivatization and polymeric samples can also be analyzed. Sample clean-up is usually much less of a problem with HPLC than GLC and biological fluids can often be directly injected onto an HPLC column. Much sample pretreatment is also avoided since aqueous solvents can be used in HPLC. A combination of these factors has meant that in the decade since its inception, HPLC has already made a significant impact in pharmaceutical, clinical, forensic and environmental analysis, and is an ideal complementary technique to GLC. Preparative HPLC is also beginning to make an impact and is likely to be more successful than preparative GLC methods with their attendant problems of volatilization and thermal instability. Already gram quantities of materials have been purified by preparative HPLC techniques. On the analytical scale, selective fluorescence and electron capture detectors capable of monitoring 10^{-10} g of an injected compound have been developed. Automatic sample injection devices allow many samples to be analyzed without operator intervention and these will clearly be useful in clinical and quality control laboratories.

To summarize, the advantages which HPLC can offer in the analysis of pharmaceutical products, body fluid samples and environmental residue samples guarantee that the wide interest generated by the technique over the last decade will be maintained during the next one.

2

Chromatographic parameters

2.1 Column capacity ratio

The column capacity ratio, k', or simply the k' value of a solute is the usual method of indicating solute retention. As shown in Fig. 2.1, k' values are obtained from the elution chromatogram by

$$k' = \frac{t_r - t_0}{t_0} \tag{2.1}$$

where t_r is the retention time of the given peak and t_0 is the retention time of the unretained (or solvent) peak. (Strictly speaking retention volumes should be used in chromatographic calculations rather than retention times. However, if constant flow rates can be assumed, retention times are more convenient to use and are obtained directly from the elution chromatogram). The use of k' values is preferred to simply quoting retention times, since the latter can vary with flow rate variations from day to day, whereas k' values remain constant.

The k' value in HPLC is related to the R_f value in thin layer chromatography (TLC) by

$$k' = \frac{1}{R_f} - 1 \tag{2.2}$$

or

$$R_f = \frac{1}{1 + k'}. \tag{2.3}$$

The R_f value in TLC is the ratio of the distance travelled by the solute to the distance travelled by the solvent front. The k' value measures the ratio of the time spent by the solute in the stationary phase to the time spent in the mobile phase.

2.2 Number of theoretical plates

The efficiency of a chromatographic column is measured by the number of theoretical plates, N, to which the column is equivalent. This parameter is

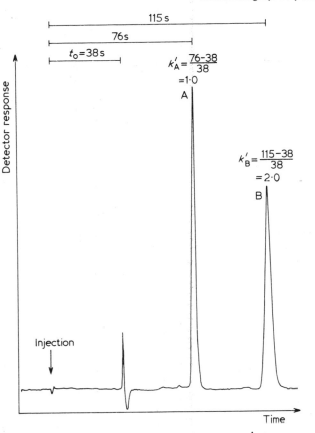

Fig. 2.1 Measurement of the column capacity ratio, k'. (The unretained peak is a refractive index effect caused by injecting the samples in pentane, with hexane as eluent.)

calculated from

$$N = 16 \left(\frac{t_r}{W}\right)^2 \tag{2.4}$$

where t_r is the retention time of the peak and W is the base width of the peak measured in the same units (mm, ml, s, etc.) and is obtained by extrapolation of tangents at the points of inflection to the baseline, as shown in Fig. 2.2. The plate number of a column is a measure of the amount of spreading of a solute band as it travels down the column, and efficient systems are characterized by high values of N. For a Gaussian peak, W represents 4 standard deviations (i.e. 4σ). It is also possible to calculate N from the width of the peak at half height, W_{HH}, using

$$N = 5.54 \left(\frac{t_r}{W_{HH}}\right)^2. \tag{2.5}$$

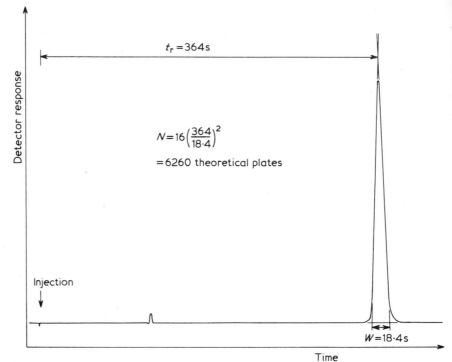

Fig. 2.2   Measurement of column efficiency, N.

2.3   Height equivalent to a theoretical plate
The height equivalent to a theoretical plate (H.E.T.P.) or simply the plate height, H, of a column, is given by

$$H = L/N \qquad (2.6)$$

where L is the column length, and it follows that with efficient columns, small values of H are obtained. H has the same dimensions as L.

2.4   Plate height curve
The plate height of a column varies with the linear velocity of the eluent, u, where

$$u = L/t_0. \qquad (2.7)$$

A plot of the dependence of H on u is known as a plate height curve and a typical example is shown in Fig. 2.3. Plate height curves provide a means of comparing the efficiency of different columns and packing materials. However, since H depends on the particle size, d_p, a series of equally well-packed columns of the same material in various size fractions will give a series of different plate height curves as shown in Fig. 2.3. To overcome this problem it is preferable to use the dimensionless parameters h and v, the reduced plate height and reduced fluid velocity respectively, where

$$h = H/d_p \qquad (2.8)$$

and

$$v = ud_p/D_m. \qquad (2.9)$$

D_m is the diffusion coefficient of the solute in the mobile phase and is estimated from the Wilke–Chang equation [14] (see Appendix 1). The reduced plate height, h, is independent of the particle diameter and measures the ratio of the plate height to the particle diameter. The reduced fluid velocity is a measure of the ratio of the rate of diffusion of solute within the particle to the rate of flow outside the particle. When h is plotted against v, the different size fractions of the same materials should in theory give identical curves, any variations found being due to differences in packing. Some experimental h vs v plots of both a chemically bonded reversed phase material and the corresponding unbonded silica are shown in Fig. 2.4. The practical implication of the minimum in the plate height curve is that maximum chromatographic efficiency (i.e. minimum value of h) will be obtained at a specific reduced velocity (and therefore at a particular operating pressure).

It can be seen from Fig. 2.4 that in well-packed columns of modern efficient packing materials, reduced plate heights of around 2 can be achieved. (For the

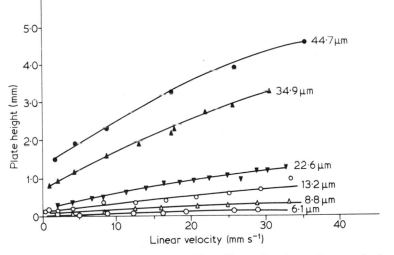

Fig. 2.3   Dependence of the plate height, H, on the eluent linear velocity, u. Column packing, LiChrosorb SI 60; particle size (column length), 6.1 μm (150 mm); 8.8 μm (150 mm); 13.2 μm (150 mm); 22.6 μm (300 mm); 34.9 μm (500 mm); 44.7 μm (1 m); column internal diameter, 2.4 mm; eluent, hexane/dichloromethane/isopropanol (90:9.9:0.125); column temperature, 23 $\pm$ 2°C. Solute, N,N-diethyl-p-aminoazobenzene ($k' = 1.2$; 1 μg injected in 1 μl). Column resistance parameters, φ', 970 $\pm$ 50. Reproduced with permission from reference 19.

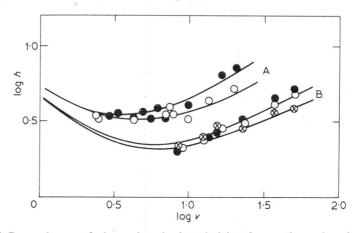

Fig. 2.4 Dependence of the reduced plate height, h, on the reduced eluent velocity, v. Column packings: (A) spherical silica (9 μm); (B) silica as in (A) (particle size 7.5 μm) containing chemically bonded short chain alkyl groups; column dimensions, 125 x 5 mm; eluents: (A) hexane/methanol (99:1); (B) methanol/water (70:30); column temperature, ambient. Solutes: (A) nitrobenzene (○, $k' = 0.3$) and p-aminoazobenzene (●, $k' = 5.1$); (B) acetone (⊗ $k' = 0.9$), 2-phenylethanol (●, $k' = 1.2$) and 2,6-xylenol (○, $k' = 1.5$). Column resistance parameters, φ', (A) 540; (B) 450. Reproduced with permission from reference 150.

bonded material, log h at the minimum was 0.33, i.e. $h = 2.14$.) Thus, for a particle diameter of 7.5 μm the plate height H was 16.1 μm or 2.14 particle diameters, and the 125 mm column generated around 7800 plates. A practical application of a 125 mm column of 5 μm silica which gave around 10 000 plates is shown later in Fig. 19.1. The column resistance parameter (see below) was 550.

2.5 Plate height equation
The first equation relating plate height to fluid velocity was given by Van Deemter *et al.* [15]. An improved version, using dimensionless terms given by Knox *et al.* [16, 17], has the form

$$h = B/v + Av^{0.3} + Cv. \tag{2.10}$$

The first term is important at very low flow rates and gives the dispersion due to axial diffusion ($B \sim 2$). The second term is a measure of the goodness of packing of the column, with $A < 1$ in the best cases. The third term predominates at high pressure and is the dispersion arising from slow mass transfer between the mobile and stationary phases. The theoretical minimum value of C for a fully porous particle is 10^{-2} [8] but it is typically between 0.05 and 0.1 for modern microparticulate materials. For materials with poor mass transfer properties h increases rapidly as v increases.

2.6 Column resistance parameter

Another term of practical importance is the column resistance parameter, ϕ', where

$$\phi' = \frac{\Delta P d_p^2}{u \eta L} \qquad (2.11)$$

where ΔP is the pressure drop down the column, u the eluent linear velocity and η the eluent viscosity. ϕ', which is dimensionless, is a measure of the resistance to eluent flow caused by the column and should have as low a value as possible. Typical values lie in the range 500–1000 for a column of porous particles. Alternatively, the column permeability K may be used, where

$$K = \frac{u \eta L}{\Delta P} \qquad (2.12)$$

Unfortunately, the experimental value of ϕ' is strongly dependent on the value chosen for the particle diameter. Since there is some uncertainty in the literature about the measurement of d_p (weight averaged or number averaged) [18, 19], h and ϕ' should always both be quoted when discussing column performance to compensate for any error in the assumed value of d_p. For example, the ϕ' values of the columns in Fig. 2.4 were both around 500. Ideally, the quantities H^2/K or $h^2 \phi'$ should be used as a measure of column performance as both terms are independent of the particle size. Low values would indicate high performance.

2.7 Resolution

The resolution, R_s, of two chromatographic peaks is a measure of their separation and is calculated from

$$R_s = 2 \frac{t_2 - t_1}{W_1 + W_2} \qquad (2.13)$$

where t_1 and t_2 are the retention distances and W_1 and W_2 the peak base widths (in the same units as t_1 and t_2) of peaks 1 and 2 respectively (Fig. 2.5). For two Gaussian peaks, effectively baseline resolution is obtained for $R_s = 1.5$. (For two adjacent isosceles triangles, baseline resolution corresponds to $R_s = 1.0$.)

By substituting for W in terms of N in equation (2.13), a general resolution equation may be derived which relates R_s to α, the column selectivity, k' and N, given by

$$R_s = \frac{1}{4} \frac{(\alpha - 1)}{\alpha} \cdot \frac{k_2'}{1 + k_2'} \cdot \sqrt{N}. \qquad (2.14)$$

The three factors contributing to R_s can be optimized independently. The selectivity, α, is defined as the ratio of the k' values of the two peaks of interest, i.e.

$$\alpha = k_2' / k_1' \qquad (2.15)$$

Clearly the two peaks can only be (at least partially) resolved for $\alpha \neq 1$, and the selectivity is optimized by changing the temperature [20] or the nature of the

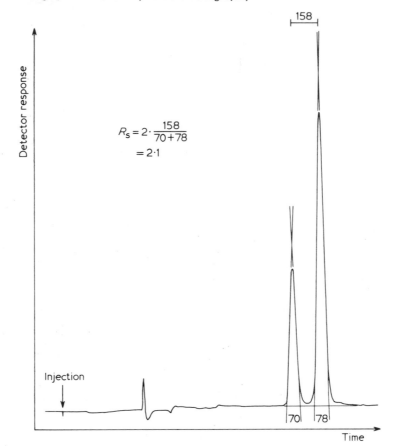

Fig. 2.5 Measurement of resolution, R_s.

mobile or stationary phases. The k' value of the solute is varied by a change in temperature or the mobile phase composition, and N can be increased by increasing the column length, decreasing the particle size or working closer to the minimum in the plate height curve. Adequate resolution for given k' and α requires a maximum plate number. As has been indicated above, typical values of N with modern HPLC columns are in the range 1000–20 000.

3

Equipment for HPLC

3.1 Complete chromatograph

In HPLC, eluent from the solvent reservoir is filtered, pressurized and pumped through the chromatographic column. A mixture of solutes injected at the top of the column is separated into components on travelling down the column and the individual solutes are monitored by the detector and recorded as peaks on a chart recorder. Therefore as shown in Fig. 3.1, the main components of a high performance liquid chromatograph are a high pressure pump, a column/injector system, and a detector. In addition, components such as solvent reservoirs, in-line filters, pressure gauges, recorders, integrators and minor components may be required. In setting up a chromatograph, the first problem to be resolved is whether to buy a complete chromatograph from a manufacturer or whether to assemble a chromatograph from modules. The disadvantage associated with a complete chromatograph is that parts which may require occasional (or frequent) attention, such as the pump or detector, are often inaccessible. In a modular equipment from a manufacturer and particularly in laboratory-

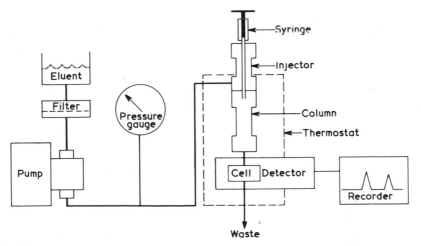

Fig. 3.1   Components of a high performance liquid chromatograph.

assembled equipment, the individual components are more easily accessible for servicing. A liquid chromatograph is in no way a 'black box' which can be expected to function and deliver answers without attention. The chromatographer must be prepared to wield a set of spanners: columns must be changed, septa if used must be replaced regularly, and columns may have to be packed or re-packed by the operator. Line filters, flow cells and pump components such as non-return valves or high pressure seals, must be removed from time to time for inspection, servicing and/or repair. Clearly, some instrumental workshop back-up is desirable, particularly in setting up a home-made chromatograph.

Since HPLC equipment has been reviewed extensively elsewhere [21—25] detailed descriptions of each part of the chromatograph will not be given here, but some comments, which may affect the results obtained with various systems, will be made.

3.2 Pumps

There are several types of pumping systems available, *viz.*, direct gas pressure, pneumatic (and hydraulic) intensifier pumps, reciprocating and screw-driven pumps. A review of pumps for HPLC is given elsewhere [26].

3.2.1 *Direct gas pressure system*

Several chromatographic systems are commercially available in which cylinder gas pressure is applied directly to the eluent in a holding coil. This system has the disadvantage that solvent changing is rather tedious as the holding coil must be thoroughly flushed out and the system is therefore not well suited to research purposes. However, it has the distinct advantages of cheapness and reliability (since there are no moving pump parts to fail) and in quality control situations where one mobile phase is in use over long periods of time, this may be the most suitable apparatus.

3.2.2 *Pneumatic intensifier (constant pressure) pumps*

A pneumatic intensifier pump is operated by gas pressure. Pressure from a gas line acts on a large area piston which drives a piston of small area. The gas pressure is thereby amplified in the ratio of the areas of the faces of the pistons (typically 20:1 up to 70:1 amplification) and high pressure liquid at a constant pressure is delivered to the system. In the event of a partial blockage in the line (e.g. in a frit or filter) constant pressure is maintained and there is a drop in the volume flow rate through the system. Commercially available constant pressure pumps deliver volumes from 2—75 ml per stroke. During the stroke little noise is registered by the detector as the flow is smooth and uninterrupted. The flow sensitivity of the detector cell will determine how much pulse damping is required in the system to suppress the detector signal caused when the flow stops during the return stroke. This problem is particularly acute in the low volume (2 ml) pneumatic pumps which recycle every few minutes. A large volume pump can usually be primed before each chromatographic run to give up to 1 h pulse-free running. Problems with pneumatic pumps usually centre around the failure of the high pressure seals and care must always be taken to use only solvents which are compatible with the material used in the manufacture of the seals. Solvents should also be well filtered to remove abrasive particles. It may prove rather tedious to change solvents with the larger pneumatic pumps because of the volumes to be flushed. Pneumatic pumps are the pumps chosen for the slurry packing of columns, as high pressure can be delivered instantly by means of an on/off valve which drives the slurry into the column. Modern constant pressure pumps are sometimes fitted with flow feed-back systems which convert them into pumping systems giving the constant flow of eluent required for accurate quantitation and reproducibility.

Hydraulic intensifier pumps work on the same principle as pneumatic intensifier pumps, but use liquid pressure rather than gas pressure on the low pressure side.

3.2.3 *Reciprocating (constant flow) pumps*

A reciprocating pump is usually driven electrically; the driven piston pressurizes either the eluent directly or hydraulic oil which, acting via a diaphragm, pressurizes the liquid from the solvent reservoir. A pulsating flow of eluent is produced (usually between 25 and 100 strokes per minute), the time-averaged flow rate being constant and depending on the pumping rate or stroke length of the piston. In the event of a partial blockage in the system, the pump attempts to deliver the required flow rate and consequently the pressure may rise dramatically. For this reason, cut-out devices, which will switch off the pump if a pre-set pressure limit is reached, are commonly used with these systems. Reciprocating pumps for analytical chromatography usually deliver up to 10 ml min^{-1}.

Faults in reciprocating pumps may be caused by wear or scratches in the non-return valves (stainless steel or sapphire spheres which must seat exactly into their holders), high pressure seals or pistons. Small pieces of grit in the non-return valves can also cause loss of pressure and for this reason a line filter should always be fitted at the inlet of such a pump. In diaphragm pumps the hydraulic oil in which the piston moves must be kept topped up. Air bubbles in the oil reservoir can cause irregular pressure variations, but these can generally be easily removed. The pulsating flow of eluent produced by a single piston reciprocating pump causes serious baseline noise with flow-sensitive detectors, and such a pump must be used in conjunction with an efficient pulse damper. One solution to the problem is the use of a twin or triple piston pump where the pistons work out of phase. This arrangement greatly reduces baseline noise.

3.2.4 *Syringe pumps*

In these pumps an electrically driven lead-screw moves a piston which pressurizes a given volume of solvent (e.g. 75—250 ml) and delivers a pulseless constant flow of solvent to the system. These pumps are expensive but reliable. A combination of two syringe pumps allows the generation of multi-shaped gradient profiles [27] and the use of flow programming.

3.2.5 General comments

Pumps must be mechanically reliable over long periods, and they are expected to deliver mobile phase at a constant flow rate and to cause as little baseline noise as possible. Constant flow rate of mobile phase is critical for the reproducibility of retention volumes and for the accurate quantitation of chromatographic peaks, and is usually achieved in modern constant pressure pumping systems by the use of flow feed-back systems. Baseline noise arising from flow can usually be reduced to manageable proportions by the use of pulse dampers. It is also desirable to be able to change quickly from one solvent to another especially in non-routine investigations. In this respect low volume pumps are more easily flushed out with the new solvent, although the system as a whole, including pulse dampers, pressure gauges etc., must be considered. In general, it is unsatisfactory and time consuming to change from one solvent to another with which it is immiscible.

A single pneumatic pump can be used with a holding coil to generate gradients [28] or alternatively, two pumps can be combined to deliver programmed amounts of two solvents. Care must be taken to use pumps which can operate successfully at very low flow rates (at the start of gradient) and which generate reproducible gradients. In general, with two pumps, low mixing volumes are required and the effects of solvent compressibility and solvent mixing on the delivered flow rate must be allowed for [27, 29–33].

3.3 Pulse dampers

In many instances, especially when operating at high detector sensitivity, the baseline noise will require damping and there are several ways of achieving this. Commercial pulse dampers utilize a coil of Bourdon tubing, or a spring-loaded diaphragm. A simple system uses a bubble of air trapped in a length of tubing to absorb the pressure fluctuations [34]. Alternatively, a column packed with glass beads has also been found useful [21]. A pressure gauge also provides some pulse damping. Data on the performance of some simple pulse dampers are available [21, 35].

3.4 Detectors

Various reviews of detection systems for HPLC are given elsewhere [36–41].

3.4.1 UV detectors

By far the most widely used detector is the UV detector which measures the change in the UV absorption as the solute passes through a flow cell (usually 10 μl volume) in a UV transparent solvent [42, 43]. UV detectors are concentration-sensitive and have the advantage that they do not destroy the solute. UV detectors can utilize the fixed emission line of a mercury lamp

(254 nm) to allow the detection of molecules with some absorption at this wavelength. By the use of a fluorescent emitter with suitable filters further detection at a range of other fixed wavelengths (e.g. 280 nm) is possible. On the other hand, the continuous emission of the deuterium lamp can be utilized in conjunction with a monochromator to provide a variable wavelength detector. The variable wavelength facility is an extremely useful way of gaining increased sensitivity in difficult analyses, as solutes can be monitored at their wavelength of maximum absorption [43–46]. The stability of current detectors is such that concentrations in the eluent of about 1 part in 10^9 of molecules with high molar extinction coefficients (ϵ in the range 10 000–20 000) can be detected. This corresponds to a few nanograms of injected sample. Double beam UV detectors are available which can record the UV spectrum of the solute if the flow is stopped while the solute passes through the cell.

Of course, not all molecules possess a sufficiently strong UV chromophore for satisfactory UV detection. Important examples include bile acids, lipids, sugars and most amino acids. One possibility is to form UV absorbing derivatives of these molecules and as discussed in the Derivatization section, this technique is often successful. Recently, bile acids [47] were detected by their end-absorption at short wavelengths (190–210 nm) and this technique could be applied to the detection of bromo-organics, sugars and steroids with isolated double bonds. Solvent purity is essential if this method is to be at all successful and sensitivities at best are no better than those of refractive index detectors. However, with some aromatics an increase in sensitivity is obtained by monitoring at short wavelength [44]. A list of the λ_{max} and ϵ_{max} values of some common chromophores is given in Appendix 2.

3.4.2 Refractive index detectors

Refractive index (RI) detectors function by measuring the change in refractive index in the eluent as the solute flows through the sample cell. To avoid drift at high sensitivity, thermostatting of sample and reference cells to better than $0.001°$ C is required. At its best, RI detection is about three orders of magnitude less sensitive than UV detection and levels of about 1 part in 10^6 (corresponding to about 1 μg injected) provide the detection limit of RI detectors. These detectors do have the advantage that many UV non-absorbing compounds can be measured directly without derivatization, and in cases where sensitivity is not important, this is clearly useful. RI detectors cannot be used in gradient elution systems since the baseline continuously varies as the changing solvent mixture causes refractive index variations. To overcome this, a step gradient system may be used where the mobile phase is changed abruptly and the baseline settles before the peaks elute. Alternatively, flow programming can be used with RI detection (p. 24).

3.4.3 Fluorimetric detectors

In a fluorimetric detector, the solute is excited by UV radiation of a given wavelength (the excitation wavelength) and the fluorescent energy which is emitted at a longer wavelength (the emission wavelength) is detected [42]. In a spectrofluorimeter, both the excitation and emission wavelengths can be selected giving increased selectivity and sensitivity. Fluorimetric detection has been successfully employed with compounds which are naturally fluorescent, or which have been reacted to form fluorescent derivatives [48, 49], and fluorimetric detection systems have given useful selectivity in trace analysis work [50]. Fluorimetric detectors should be inherently more sensitive than UV detectors, and although the sensitivities of commercial instruments of the two types have been similar until now, the latest generation of fluorimetric detectors appear to have increased sensitivity.

3.4.4 Electrochemical detectors

Several designs for electrochemical detectors have been published [51–53]. Their sensitivity is comparable with that of the UV detector, i.e. 1 part in 10^9 to 1 part in 10^{10}, and for complex problems, electrochemical detection may provide useful selectivity for electroactive compounds. Problems with these detectors involve contamination of the electrodes by adsorbed impurities. A polarographic detector with a movable electrode was used to study peak profiles eluting from a chromatographic column [54], and a capacitance detector for HPLC has also been described [55].

3.4.5 Electron capture detectors

The electron capture detector [56] is useful for detecting halogenated molecules and a commercially-available electron capture detector for HPLC has recently appeared [57]. Column effluent (usually from an adsorption column, e.g. using a hexane mobile phase) is volatilized in an oven purged with nitrogen and halogenated solutes are detected. The detector is particularly useful in the analysis of chlorinated insecticides and polychlorinated biphenyls, and has recently been used to detect DDT in milk [58]. Very low levels of suitable compounds have been monitored with this detector, for example, 4×10^{-11} g of aldrin [59].

3.4.6 Reaction detectors

A potentially useful and selective range of detection systems can be envisaged involving reactions of the solute(s) after elution from the column. Such a system must employ a very fast chemical reaction otherwise the solute will become unacceptably diluted by diffusion; likewise tubing and mixing chambers must be strictly limited in volume [60]. Several useful post-column reaction detectors include the well known ninhydrin reactor for amino acids [61], the 2,4-dinitro-phenylhydrazine (2,4-DNP) reactor for carbonyl compounds [62] and the cerium detector for carbohydrate analysis [63]. Other examples are given in Section 4.3.

3.4.7 Transport detectors

In the transport detector, column effluent is picked up on a wire or porous disc, the solvent evaporated and the solute samples pyrolyzed in a hydrogen/air atmosphere, and detected using a flame ionization detector [64–66]. The transport detector is a fairly universal mass detector for HPLC and has been particularly useful in the analysis of lipids. It suffers, however, from a rather poor signal to noise ratio. This arises from the motion of the wire winding and unwinding round two drums and passing through ovens etc., and also from the low efficiency of sample transfer onto the wire. 'Memory' effects can also occur if the wire is not properly cleaned between analyses. Its detection limit is comparable with that of the RI detector.

3.4.8 HPLC–mass spectrometric interfaces

The idea of an HPLC–MS combination, although expensive, is undoubtedly attractive. The technique involves transporting the eluent (or a fraction thereof) into a chamber where solvent is evaporated and solutes are ionized and detected in a mass spectrometer. Various methods of stream splitting, transport and solute ionization have been reported [67–74]. A serious limitation is that involatile buffer solutions cannot be used for separations monitored by these systems. One of the systems is now commercially available [73].

3.4.9 Miscellaneous detection systems

Various other detectors for HPLC have been described, including a commercially available microwave plasma detector [75], and infrared (IR) detector [76, 77], as well as atomic absorption [78, 79] and flame photometric [80] detectors for metals, a heat of adsorption detector [40, 81], a spray impact detector [82], a phosphorus/sulphur detector [83], a dual UV-fluorescence detector [84] and radiometric detection systems [39, 85, 86].

3.5 Columns and injection systems

The heart of the chromatograph is the column–injector combination. This is the critical area where efficiency can be easily lost. Important considerations are the column dimensions, how the column packing is terminated top and bottom, the type of connection from the column to the detector and the type of injection system.

3.5.1 Column dimensions

As shown by Knox and Parcher [87] the most efficient chromatography can be expected if the solute, introduced centrally at the top of the column as a point injection, does not reach the walls of the column as it travels down the column. At the walls there is a region of disturbed packing which can reduce chromatographic efficiency. The relationship between column length and diameter and particle diameter for these so-called 'infinite diameter' columns packed with porous particles and taking into account the disturbed wall region has been given by Knox, Laird and Raven [54] as

$$\frac{(d_c - 60d_p)^2}{Ld_p} \geqslant 16\left(\frac{1.8}{\nu} + 0.060\right) \tag{3.1}$$

where d_c is the column diameter and L is the column length. In practice, infinite diameter columns may have the dimensions shown in Table 3.1. Injections into infinite diameter columns are normally made by syringe.

Table 3.1   *Maximum column lengths (mm) for infinite diameter effect*

Column diameter (mm)	Injection volume (μl)	Particle size		
		5 μm	10 μm	20 μm
5	10	300	220	140
	1	470	340	220
7	10	920	670	440
	1	1130	800	530
10	10	2350	1700	1100
	1	2500	1800	1180

Data taken from reference [88] with permission.

For difficult separations, two or more columns can be coupled together to give a corresponding increase in the number of plates [19] and with extremely complex mixtures capillary HPLC columns which have yielded up to 50 000 [89] and even 250 000 plates [90] may be the answer. In the separation of two poorly-resolved components the effective length of the column may be increased by the use of recycle chromatography [91, 92] in which the column effluent is fed back to the pump and continuously re-chromatographed on the same column. If recycle chromatography is to be successful the extra-column dead volume must be minimized.

3.5.2   Column connections
It cannot be over-emphasized that the volume of connecting tubing between the injection system and column and the column and detector must be kept as low as possible. In the first case the problem can be overcome by designing the injector and column as an integral unit, thereby minimizing dead volume. Even in some modern chromatographs, however, the column and detector are connected by an excessive length of (not always narrow bore) tubing. Ideally not more than 200 mm of 0.25 mm i.d. (internal diameter) stainless steel tubing should connect the column to the detector inlet (also 0.25 mm i.d.) via a zero dead volume ('drilled out') coupling. Corresponding maximum lengths for other bores are: 15 mm for 0.5 mm i.d. and 3 m for 0.125 mm i.d.

3.5.3 Injection devices

(a) Syringe

As already mentioned, syringe injection (in combination with infinite diameter columns) gives the most efficient chromatographic results [54, 93, 94]. In general, several research groups have found that the most efficient configuration is to inject through a septum into a small bed of glass beads, glass wool or a teflon disc above the column and separated from the column packing material by a wire gauze [93, 95, 96]. Best results are obtained if the needle length is arranged such that injection occurs just above the gauze. A representative syringe injection system is shown in Fig. 3.2. The advantage of this system is that pieces of septum do not reach the packed bed, the syringe needle does not become blocked with particles of column packing material and the beads or wool are easily replaced without disturbing the packed bed. Also, the gauze can be easily removed to inspect the top of the column whenever a drastic fall in efficiency suggests that the packing may have settled in the column. The original performance may be recovered by topping up the column

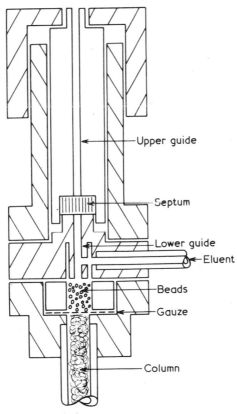

Fig. 3.2   Syringe injection system for HPLC.

with a thick slurry of packing material and levelling off at the required height. These columns have obvious advantages over columns where there is a fixed frit at the top of the column (often introduced to facilitate transport of pre-packed columns). In this case, loss of efficiency by settling of the packing material cannot be easily identified or remedied. Other groups have recommended injecting directly into the column packing material [94, 97].

Problems with syringe injection include the limitation of the operating pressure to around 1000 psi (6.8 MPa) to avoid fracturing the syringe barrel, the blocking of needles with septum material and the short life of the septum in many solvents; an advantage is that very little of the solution is wasted.

(b) Injection valves

As shown in Fig. 3.3 injection valves operate by loading a sample loop while the mobile phase is pumped via a by-pass system to the column [26]. Valves may have either an internal fixed volume loop or an external replaceable loop. By means of a lever the flow is switched through the loop, carrying with it the sample solution onto the column. Point injection is impossible to achieve by this means, although the use of 'curtain flow' may help to approximate [93]. On the other hand, very reproducible injections are possible and this is clearly of importance in quantitative work. Injection valves tend to be rather wasteful of sample solution, as excess solution is required to ensure that the loop is completely filled. The critical area of injection valves is the engineering of the various internal rotating interfaces and connecting tubes, to ensure that all the channels are well swept out by the eluent. External loop injectors are convenient from the point of view that the loop volume is easily changed by replacing the loop, and large volumes can conveniently be injected against pressures of up to 7000 psi (48MPa).

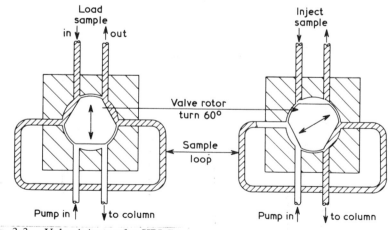

Fig. 3.3   Valve injector for HPLC.

3.5.4 Automated injection devices

Devices are available commercially which will automatically inject up to about 100 samples. These devices clearly are very useful in the clinical and industrial field where large numbers of samples have to be analyzed.

4

The practice of HPLC

4.1 General comments

4.1.1 The mobile phase

In gas–liquid chromatography selectivity is achieved by having available a wide range of columns packed with different stationary phases. In HPLC, rather the reverse is true. By having available a limited number of supports (e.g. silica and alumina, one or two reversed phase materials, possibly a polar bonded material, and possibly a cation and anion exchanger) selectivity is achieved by variation of the column temperature or the composition of the mobile phase (e.g. by the use of more than one solvent, the addition of organic solvents such as methanol or acetonitrile to buffer solutions, variation of the pH and ionic strength of buffer solutions, or the addition of small amounts of amines, acids, buffers, electrolytes or detergents to mobile phases).

4.1.2 Gradient elution and flow programming

HPLC columns may be run isocratically, i.e. with a constant composition eluent or they may be run in the gradient elution mode in which the mobile phase composition varies during the run. Alternatively, there may be a pH or ionic strength variation throughout the gradient programme. Gradient elution is a means of overcoming the general elution problem [98] which is the problem of dealing with a complex mixture of solutes, some of which elute quickly while others have high k' values. It is the analogue of temperature programming in GLC. Gradient elution techniques are also useful as a means of obtaining resolution in complex separations. A gradient device is undoubtedly useful when 'scouting' for suitable solvent conditions to chromatograph an unknown mixture. A disadvantage, however, is that time is lost in reconditioning the column for a subsequent run. Other problems encountered during a gradient run include a drifting baseline due to the slightly different UV absorbances of the different solvents, the appearance of spurious peaks due to solvent impurities [99] and flow rate variations due to changes in liquid compressibilities.

An alternative technique known as flow programming [100–102] may be employed in cases where gradient elution is ruled out (e.g. with refractive index detectors, or in liquid—liquid partition chromatography). In this case, the flow rate of a constant composition mobile phase is increased during the run.

4.1.3 Operating temperature

HPLC columns are normally run at ambient temperature, although increased efficiency can be obtained at higher temperatures due to the lowering of the eluent viscosities and the corresponding increase of the diffusion rates of solutes in the mobile and stationary phases. Mobile phases should be chosen with as low a viscosity as possible (e.g. methanol is preferred to ethanol in reversed phase chromatography). There are problems of thermostatting the column and detector at higher temperatures if baseline noise and drift are to be avoided and these must be balanced against any advantages. Elevated temperatures have been found useful in ion exchange chromatography where efficiencies are often poor. From the point of view of day-to-day reproducibility of chromatographic data, it is an advantage to have the column well thermostatted at or slightly above ambient temperature.

4.1.4 Identification of solute peaks

Solute peaks in an elution chromatogram are usually identified by their having the same k' values as standards. However, it is unwise to identify a solute peak on the basis of only one such experiment, since frequently several compounds can elute with the same k' value, or one peak may be made up of several components. By varying the mobile phase and hence k' values of the peak of interest and the standard peak, more convincing evidence may be obtained. Alternatively, the sample and the standard may be chromatographed on a different packing material. A solution of the standard added to the unknown solution should cause no distortion of the shape of the unknown peak if the two are identical. Ideally, the solute should be collected from the column and independently identified. The UV spectrum of the solution may be obtained or a micropellet prepared for (infrared) IR analysis. Off-line mass spectrometry is usually the most powerful means of solute identification.

For solutes which undergo enzyme reactions (such as nucleotides), the enzyme shift technique can be a useful, specific method of peak identification. For example, the enzyme hexokinase quantitatively converts ATP to ADP and hence both peaks can be positively assigned. Radioactive standards can also be used as a confirmatory test of peak identity. In this case, the radioactivity of effluent fractions is monitored.

4.2 Quantitative HPLC analysis

Quantitative analysis is carried out by measuring the areas or heights of the peaks in the elution chromatogram. This can be done either manually or electronically. The measurement of peak areas is an inherently more accurate method of quantitation than peak height measurement, but there are occasions when the latter method is preferred. For example, in trace analysis problems peak height measurements are used, since peak heights are less affected by the presence of interfering peaks than are peak areas. Kirkland has shown that for two overlapping peaks (with resolution $R_s = 1$), the relative peak heights can be varied from 32/1 to 1/32 with an accuracy of better than ±3 per cent, whereas for the same resolution and accuracy, peak areas cannot be changed by more than from 3/1 to 1/3 [103]. A further point is that peak area measurement is much more affected by flow rate variations than is peak height measurement [104]. Therefore, in quantitative analysis in general, and particularly when using peak areas, the pumping system must be capable of delivering a constant flow rate or the signal must be recorded *vs* volume rather than time. Peak height reproducibility, on the other hand, requires a reproducible injection distribution and this is often difficult to achieve with on-column syringe injection. Provided a calibration plot of peak height or peak area *vs* amount injected is linear over the required range, either method is satisfactory.

Measurement of peak heights is straightforward. Peak areas are best measured by multiplying the peak height by the peak width at half height. This gives a precision of around 3 per cent.* Measurement of peak area by computing ½ x base width x peak height gives a precision of around 4 per cent. In both cases a sufficiently fast chart speed should be used to ensure that peak widths can be measured accurately. Other less common procedures involve counting the small squares under a peak on the chart paper. This method is tedious but is useful for broad low peaks. Alternatively, peaks can be cut out and weighed, but this destroys the chromatogram. (A photocopy of the chromatogram can also be cut out and weighed.) A planimeter can also be used to measure the area.

A description of the electronic integrators available for peak area measurement is beyond the scope of the present text. Suffice it to say that computing integrators are available which will print out results in a report form, giving a table of results such as the percentages of each component in the mixture, the ratio of two components, the amount of an impurity present etc.

There are several points to remember when performing a quantitative analysis. Firstly, since the common HPLC detectors respond differently to different solutes, a response factor must be determined for each component in a mixture before the peak area (or height) can be converted into the amount of each particular solute. Response factors must be calculated individually for each component by injecting standard amounts of the pure component and measuring the peak area. Internal and external standards can also be used to calibrate the system for quantitative analysis. An internal standard is a compound which is added in known amounts to the mixture to be analyzed as well as to the standard solutions used to calibrate the system. Quantitation is carried out by comparing

*Precision values are quoted in terms of the coefficient of variation which is the standard deviation divided by the mean expressed as a percentage.

the solute peak area with that of the standard peak area. The k' value of the internal standard must be chosen such that it does not interfere with any of the other peaks. The method has the advantage that any loss of sample solution (e.g. on syringe injection against relatively high pressures) does not affect the analysis since both solutes of interest plus internal standard are lost proportionately. Internal standards may also be added to body fluids before the extraction steps are carried out. If the standard has a similar chemical constitution to the compound of interest to be extracted, losses of both the standard and the component of interest should be the same throughout the procedure.

External standards can also be employed. This method involves alternate analysis of standard solutions and of the samples containing the component of interest. From the injections of standard solutions a plot of peak area *vs* amount injected may be constructed and used to quantitate the peak areas from the sample solutions. If analyses are carried out in duplicate or triplicate, much of the error involved in injecting sample and standard consecutively can be eliminated.

4.2.1 Accuracy and precision in HPLC
The accuracy of an analysis is the ability of the method to measure the quantity being determined. Accuracy can be checked by weighing the components of a mixture and comparing the HPLC result for the amount present with the expected result. Precision on the other hand is the ability of the method to give the same result in a series of replicate determinations. The accuracy of the method is governed by the ability to calibrate the system using standards as described above. Precision depends on the method used for quantitation. The manual methods described above have precisions in the region of 3—4 per cent while electronic integration can give a precision of 0.5—1 per cent. (The reproducibility of a method is governed by the ability of independent analysts in different laboratories to give the same result using the same method). Scott and Reese have recently discussed the effect of temperature and solvent composition on the precision of HPLC measurements and the effect of solute mass on peak retention times and column efficiencies [105].

4.3 Derivatization
Derivatization in HPLC is carried out to enhance the detectability of various classes of compounds for which sensitive detectors are not at present available. Most work has been done on the labelling of compounds with chromophores and fluorophores for detection using UV spectrometers and fluorimeters respectively. Many examples of derivatization are given in the Applications sections of the book and only some general points will be made here.

Either pre-column derivatization or post-column derivatization can be carried out. With pre-column derivatization, there are no restrictions on the solvents, reagents or reaction rates chosen, and excess reagents can be removed prior to injection. The danger of artefact formation is present and should be checked by positive identification of the eluted peaks. In the derivatization of, for example, a triketone with more than one functional group capable of being derivatized there is the possiblity of a range of derivatives being formed from one solute. It is clearly necessary to check that the derivatization reactions are quantitative, or that sample derivatizations proceed in an analogous manner to derivatizations of standards. Pre-column derivatization can also be useful in achieving selectivity in trace analysis [49, 103].

Examples of pre-column derivatization to form UV chromophores include the treatment of ketosteroids with 2,4-DNP [106, 107], the benzoylation of hydroxysteroids [108] and the esterification of fatty acids [109–112]. Fluorophores have been introduced into amino acids, biogenic amines and alkaloids by treatment with dansyl chloride (5-dimethylaminonaphthalene-1-sulphonyl chloride) [48, 49]. Several publications have appeared on the HPLC separations of PTH (phenylthiohydantoin) amino acid derivatives [113–120]. The hydrolysis of organophosphorus pesticides to phenols and carbamates to amines followed by dansylation of the phenols and amines is discussed later (Sections 16.2.1b and 16.2.1c).

The relative retentions of compounds change, of course, on derivatization, as was noted in the chromatography of ketosteroids [107]. The epimeric forms of some vitamin D_3 metabolites were resolved as their trimethylsilyl derivatives, whereas the underivatized epimers were not separated [121]. Reaction of ionic sites such as carboxylic acid groups may result in more efficient chromatography, but on the other hand, derivatization tends to make all the solutes more alike and hence potentially more difficult to separate.

Less common methods of derivatization include the pairing of ionic solutes with UV-absorbing or fluorescent counter ions (see Section 6.9.2), (which has the added advantage that the original solutes are easily recovered), the introduction of radioactive groups such as (^{14}C) dansyl chloride [122] and the introduction of groups with easily identifiable mass spectral patterns such as dansyl derivatives of biogenic amines [123].

Post-column derivatization is carried out on the separated solutes as they emerge from the chromatographic column. In HPLC, this places severe restrictions on the derivatization reactions since dilution of the eluted peak must be minimized. This means that very fast reactions must be employed and the reagents and mobile phase must be compatible. With post-column reactors, artefact formation is less likely.

Examples of post-column reactions for use with UV detectors include the reaction of amino acids with ninhydrin [61] and fluorescamine [124], fatty acids with o-nitrophenol [125], ketones with 2,4-DNP [62], and the thermal [126] or acid/phenol treatment [127] of carbohydrates. An oxidation detector for the fluorimetric analysis of carbohydrates in body fluids using Ce(III) fluorescence has been reported [63] as has the analysis of thioridazine and metabolites in body fluids by oxidation to fluorescent products [50].

The use of derivatization techniques in TLC and HPLC has been covered in depth in a recent text by Lawrence and Frei [49].

4.4 Column packing materials

4.4.1 Adsorbents
The earliest types of HPLC packing materials were large diameter fully porous adsorbents (>30 μm) such as LiChrosorb (Merck) and the Porasils (Waters). The large diffusion distances within these particles meant that the materials had poor mass transfer characteristics. An improvement in their mass transfer properties was achieved with the preparation of pellicular supports, where a thin layer (1–2 μm) of silica or alumina was coated onto the surface of glass beads (also called superficially porous, or controlled surface porosity supports). Examples of pellicular supports include Zipax (Du Pont), Corasil I and II (Waters), Vydac (Separations Group), Pellosil (Whatman) and Perisorb A (Merck). These supports have low surface areas and are normally used for liquid–liquid partition chromatography, although some can also be used for adsorption chromatography.

However, pellicular materials had low sample capacity, as most of the column was filled with glass beads and currently, the most efficient HPLC packings are based on fully porous particles with diameters of around 5 μm. These microparticulate materials combine higher capacity with short diffusion distances and probably represent the optimum practical compromise between efficiency, ease of packing and operating pressures. It is also desirable for the particles to have fairly wide pores, i.e. >5 nm, otherwise solute molecules can become trapped in the small pores thereby giving rise to tailed peaks. The two most commonly used adsorbents are silica and alumina, although the use of other metal oxides such as ceria, zirconia or thoria may provide packing materials with more alkali resistance than either silica or alumina. Examples of commercially available adsorbents are given in Appendix 3. A comprehensive list of the various types of packing materials available for HPLC has been compiled by Majors [128].

4.4.2 Polymer-coated supports
In 1967, Horvath, Preiss and Lipsky coated the surface of glass beads with a polystyrene/divinylbenzene polymer [11]. Sulphonation of the resulting polymer produced a layer of cation exchange resin over the glass bead surface, and similarly chloromethylation followed by reaction with a tertiary amine produced an anion exchange resin. Previously, ion exchange materials had been based on synthetic cross-linked polymers. These materials had the disadvantage that they changed their volumes in organic solvents and under different pH and ionic strength conditions. They also deformed at high pressure and had poor mass transfer characteristics. The new pellicular materials had good mechanical stability and the shorter diffusion distances led to better efficiencies than had

been hitherto achieved, for example in the analysis of nucleotides [11]. A range of porous layer beads coated with ion exchange resins is now commercially available and includes Zipax SCX and SAX (Du Pont), AS Pellionex SAX (Whatman) and Pellicular Anion (Varian). Glass beads (or superficially porous glass beads) could also be coated with hydrocarbon or polyamide [129] polymers and examples of each type are collected in Appendix 4.

4.4.3 Chemically bonded stationary phases

Halasz and Sebestian [12] esterified the silanol (Si–OH) groups on the surface of silica with an alcohol and used the resultant phase in HPLC. The Si–O–C bonds so formed, however, were hydrolytically unstable. Conversion of the silanol groups to Si–Cl groups using thionyl chloride, followed by reaction of the silica chloride with amines, gave a range of Si–N bonded phases [13] which were hydrolytically stable in the pH range 4–8. Other bonded phases can be prepared by reaction of silica chloride with Grignard or organolithium reagents, the resulting Si–C bonds having good hydrolytic stability [130–135].

However, the commercially available bonded materials are based on reactions of organochlorosilanes or organoalkoxysilanes with surface silanol groups. Two types of reaction are involved. Firstly, a chloro- or alkoxysilane is hydrolyzed and partially polymerized and the resulting silicone polymer chemically bonded to the silica surface via hydrolytically stable siloxane (Si–O–Si) bonds. The Permaphase supports (Du Pont) are formed by bonding polymers to the surface of Zipax [136, 137]. Alternatively, direct reaction of the chloro- or alkoxysilane with the anhydrous support results in the formation of monomolecular siloxane bonds. Bondapak C_{18} (Waters), Zorbax ODS (Du Pont), Vydac RP (Separations Group), Partisil-ODS (Whatman), Spherisorb-ODS (Phase Separations), LiChrosorb RP-8 (Merck) and ODS- Hypersil (Shandon Southern) are prepared by this process. References to the preparation and applications of chemically bonded phases are contained in review articles [128, 138–143] and in recent publications [96, 144–159].

The most common substituents so far employed in HPLC have been hydrocarbons, especially octadecylsilyl (ODS) groups i.e. $C_{18}H_{37}Si-$, cation exchangers ($-Si(CH_2)_n-C_6H_4 \cdot SO_3H$), anion exchangers ($-Si(CH_2)_nNH_2$ or $-Si(CH_2)_nNR_3^+X^-$) and polar groups such as cyanopropyl or the bonded ether phase, γ-glycidoxypropyl. Although the range of potential bonded phases is unlimited, the useful range of such phases should hopefully comprise at most, two bonded hydrocarbon phases (a long and a short chain material), a cation and anion exchanger, and one or two polar bonded phases. The reason that the list is so short is, of course, that a wide variation in selectivity can be obtained by variation of the composition of the mobile phase (see Section 5.3.).

Modern HPLC supports with organic groups monomolecularly bonded to fully porous 5 μm particles give high speed analyses with efficiencies similar to those obtained with the unbonded silica support (Fig. 2.4.); mass transfer is fast because the solutes do not have to diffuse into a polymer bead or polymeric coating on a glass bead [160]. The range of bonded phases available has greatly increased the scope of HPLC for analyzing a wide range of compounds. Examples of commercially available chemically bonded phases are given in Appendix 5.

4.5 Choice of support material for a separation problem

When choosing a packing material to attempt a separation, it is extremely useful if the formulae of the solutes are known. If related molecules, or molecules in the same class have been separated, this system may be taken as a reasonable starting point. The most useful pieces of information when performing an unknown separation are;

(i) the molecular formula, or presence of known functional groups, especially basic or acidic groups;

(ii) the solubility of the molecules in organic solvents and in water, acids or bases;

(iii) the UV spectrum (assuming a UV detector is being used) to ensure that detection is performed where the molecule has reasonable absorbance;

(iv) the presence of impurities — it is essential to chromatograph and identify the solute of interest and not an artefact or impurity;

(v) TLC information (see below).

Non polar and moderately polar compounds are normally separated by adsorption chromatography or reversed phase chromatography and highly polar molecules by reversed phase chromatography. Acids and bases are chromatographed by anion and cation exchange respectively, or by ion pair or soap chromatography. Polymers are separated by exclusion chromatography. Usually more than one system can be developed for most separations, hence no very detailed general scheme will be given here — most of the classes of compound likely to be encountered are mentioned in the Applications sections. An important practical point in developing an HPLC method is to choose a system where the k' value of the solute of interest can be varied easily. For example, in ion exchange chromatography this may be accomplished by variation of the pH of the mobile phase and in reversed phase chromatography by variation in the components of the mobile phase (e.g. the methanol/water ratio). This flexibility allows the solute to be chromatographed free from interferences which vary from sample to sample.

With very complex mixtures, a preliminary size separation can be carried out by exclusion chromatography and the individual size fractions further analyzed by HPLC.

4.5.1 TLC and HPLC

When developing an HPLC separation it is useful to consider any TLC information which might be available. Using the TLC adsorbent type (i.e. silica or alumina) in the HPLC system and assuming the TLC solvents to be compatible with the HPLC detector, a reasonable starting point is to use an HPLC mobile phase of lesser polarity (since TLC plates are generally more active than

HPLC columns). One advantage of HPLC over TLC is that in HPLC the full resolving power of the column is brought to bear on all solutes, as all solutes pass down the entire length of the column whereas in TLC, poor resolution is obtained at low R_f values. One rather inconvenient disadvantage of HPLC, not shared by TLC, is that compounds which adsorb strongly to the packing material and do not elute, are not detected visually as they can be on a TLC plate. Further information on sample structure in TLC may be obtained by the use of spray reagents. Another advantage of TLC is that multiple spottings on a single broad plate do not significantly increase the analysis time, whereas in HPLC, all the compounds from one run must usually elute before the next sample is injected.

4.6 Column packing techniques

The chromatographer is faced with the problem of whether to pack his own columns or whether to buy pre-packed columns from the manufacturers. The advantages of the former course include ultimate financial saving, and the convenience of being able to pack a column quickly whenever the need arises.

There are two distinct approaches to column packing depending on particle size. For relatively large particles (>15–20 μm) a dry packing technique can be used and involves adding the dry column packing material slowly, either continuously or in small aliquots, while simultaneously bouncing, tapping and rotating the column [161], or by a modified tap-fill method [162].

For small particles, this technique is not feasible and a slurry packing method is used. This involves coupling the column to be packed to a reservoir, filling the column with supporting liquid and the reservoir with a slurry of the column packing material in a suitable supporting liquid. The reservoir/column assembly is connected to a pneumatic intensifier pump and a pressure of 3000–10 000 psi (20–68 MPa) is instantaneously applied to the system, thus forcing the slurry into the analytical column. Particle fractionation should be prevented during slurry packing and any agglomerates present should be broken up. Fractionation is prevented by rapidly packing the slurry into the column at high pressure and agglomerates can be dispersed by using the correct slurry liquid and by agitating the slurry in an ultrasonic bath prior to packing.

In general, due to their different polarities, adsorbents and bonded phase materials require different slurry liquids. Silica and alumina have been packed using balanced density slurries, i.e. with liquids such as methyl iodide and 1,2-dibromoethane which have densities similar to the adsorbent and which therefore prevent fractionation of the adsorbent [136, 163–165]. High viscosity liquids have also been used [166] and Kirkland found a dilute aqueous ammonia solution useful in charging the surface of silica particles, thereby reducing agglomeration [167]. Carbon tetrachloride slurries have also been successfully used [93, 94]. Slurries are usually fairly dense (up to 30 per cent w/v) [93, 163] although dilute slurries (1–5 per cent w/v in methanol) have also been successfully packed [168]. (Organobromine and iodine solvents are not recommended for bonded phase slurry packing as any halogen present could cleave off the bonded groups.)

4.7 Trace analysis

In trace analysis, small amounts of the compound of interest must be analyzed in the presence of a large amount of interfering matrix. Typical trace analysis problems which have been dealt with by HPLC include the analysis of drug metabolites in body fluids and the analysis of pesticide residues in environmental samples. Success in the difficult area of trace analysis requires paying attention to several factors. Chief among these are;

(1) obtaining maximum resolution;
(2) a suitable calibration technique;
(3) choice of equipment;
(4) sample pretreatment.

The achievement of maximum resolution has already been mentioned in connection with the resolution equation (2.14). For reasons discussed in Section 4.2, quantitation in trace analysis is by the peak height method. This method is more accurate in trace analysis work and gives results of acceptable precision (see p. 27 for the distinction between accuracy and precision). Thus if possible, the operating conditions are chosen such that the solute of interest has a k' value in the range 0.5–1.5, with a correspondingly sharp peak. However, this may be feasible only when using a selective detector, and with, for example, a UV detector, larger k' values may be required in order to separate the trace solute from interfering peaks.

The equipment chosen also plays a major role in trace analysis work. Since maximum sensitivity is used, the pump/pulse damper system capable of producing the least baseline noise will contribute significantly to the success of the analysis. Also, selective detection systems may be invaluable in reducing interferences from the matrix. Examples of selective monitors used in trace analysis include the use of fluorimetric detectors for the analysis of thioridazine and its metabolites in body fluids [50], electrochemical detectors for body fluid analysis [52], reaction detectors for the analysis of carbohydrates in body fluids [63, 126, 127] and an electron capture detector for the analysis of chlorinated pesticides in milk [58].

Stop flow injection by syringe can be used to load a large volume of solution onto the column, and at higher pressures injection valves can be used. As discussed in detail by Karger and co-workers [169], injection of a large volume of a dilute solution will lead to more efficient chromatography than injection of smaller volumes of more concentrated solutions. This arises because volume overload effects are less serious than mass overload effects. Thus to increase the sensitivity of the method, increasing volumes should be injected until chromatographic performance begins to deteriorate badly. Kirkland has suggested injecting up to one third of the baseline volume of the trace component peak [103]. Because of the low capacity of porous layer beads, fully porous materials are required in trace analysis to cope with the large amounts of sample injected.

Two additional techniques, column switching and column backflushing, have been developed to deal with the wide range of compounds encountered simultaneously in trace analysis. In column switching, a relatively large volume of sample is injected onto a short analytical column A (Fig. 4.1). The early eluting trace component of interest passes into the longer analytical column B which has the resolving power to separate the trace component from any early eluting interferences. These components are held at the top of column B and the valve switched to divert later components eluting from column A directly to the detector. After passage of these compounds through the detector, the flow is again diverted through column B and the desired separation of trace component from interferences achieved. The time required to optimize a column switching method probably restricts its use to routine separations.

Column backflushing is a method of removing strongly held components from the top of a column after the components of interest have been analyzed. As shown in Fig. 4.2., the sample is injected at the top of the column and after

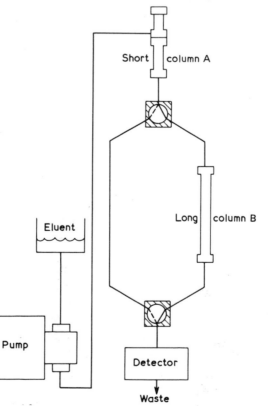

Fig. 4.1   Column switching.

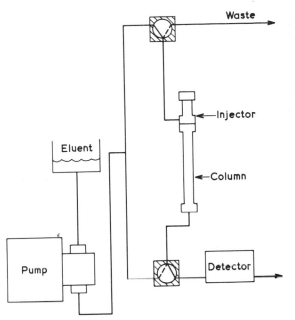

Fig. 4.2   Column backflushing.

elution of the solute of interest the switching valves are used to lead the eluent into the bottom of the column and out of the top to waste. Thus strongly held components are fairly quickly stripped from the top of the column. The critical features of both techniques are that the valves should withstand the operating pressures, and should not degrade peak shape by dilution.

Finally, sample pretreatment is an important means of reducing interference in trace analysis. Several methods are available and include extraction of the solute, making use of solubility differences between the solute and the interferences, or the presence of cationic or anionic groups in the solute molecule. Derivatization may also be used to increase the selectivity of the detection system to the trace solute. Chromatographic techniques can also be effective for the prior clean-up of samples. Size separations can be carried out by exclusion chromatography, ionic compounds can be purified by ion exchange chromatography and relatively non polar compounds can be concentrated by passage down an adsorbent column. Organic compounds in river samples were concentrated at the top of a reversed phase column before being eluted with an appropriate solvent, the technique being termed trace enrichment analysis [170].

Kirkland has published a comprehensive review on trace analysis [103].

4.8   Capacity of a chromatographic support
The linear capacity of an adsorbent has conventionally been taken as the amount of solute which may be applied to the column before a 10 per cent loss in

efficiency was observed [7, 18, 166, 171–173]. This occurred typically for $10^{-3}-10^{-4}$ g of solute per gram of support. The parameter has obvious practical usefulness especially in relation to preparative HPLC. However, Done [174] has recently shown, using very efficient columns (reduced plate heights of 2–3), that the efficiency (plate height) is linearly dependent on sample load down to very low sample loads (1 μg of sample per gram of support). As a measure of adsorbent capacity, Done has proposed the term 'relative capacity', i.e. 'the sample weight per gram of adsorbent which increases the H.E.T.P. of the column by 0.1 mm from the minimum value' [174]. For analytical columns, the minimum amount of solute dissolved in less than 5 μl of solvent should give the best results. Linear capacity has also been measured by the amount of solute required to cause a 10 per cent decrease in k' [175]. However, it was shown that a substantial fall in efficiency on the very efficient columns was not accompanied by any corresponding decrease in k' value [174]. Column capacity was proportional to the surface area of the adsorbent, and depended markedly on the mobile phase and the k' value of the solute, with relative capacity values being inversely proportional to log k' [174]. When measuring column capacities using, for example, a UV detector, care must be taken when chromatographing large weights of solutes that the detector is still responding linearly. Otherwise poor peak shape can arise as a result of detector problems rather than column overloading. One solution is to monitor at a wavelength where the solutes have little absorbance.

4.9 Preparative HPLC

The majority of HPLC users will at some time wish to collect samples from the HPLC column for further investigation. The amount of sample required may vary from a few milligrams for structural identification by MS, NMR, IR and elemental analysis up to gram quantities or more for further synthetic work. Before attempting a preparative separation, an analytical system should be developed with good resolution and convenient k' values (1–3) of the peaks to be collected. By scaling up the column length and width, and the particle diameter, the same separation can be achieved in the same time for the same pressure drop [176]. The sample load can then be increased until the resolution of the desired components is slightly better than adequate, and the flow rate increased until the separation is just adequate [100]. This ensures maximum throughput with the preparative column. In contrast to the analytical plate height curve which has a distinct minimum (Fig. 2.4), in preparative work the overloaded plate height curve is fairly flat [174]. Hence there is much less penalty incurred (in terms of plate height) in working at high flow rates [100, 101, 174]. This factor also helps to achieve a higher throughput in preparative HPLC. Good analytical efficiencies are achieved on wide bore, preparative columns [177–180].

Regarding equipment requirements for preparative work, a pump is required which is capable of delivering high flow rates (up to 100 ml min^{-1}). Longer and/or wider columns are obviously necessary, the dimensions being determined by the quantity of material to be separated and the resolution required. Injection systems capable of coping with millilitre injections are required. These include loop injectors where the loop size can be varied to provide variable injection volumes; syringe injectors using a stop flow technique; or alternatively, the chromatographic pump can be used to inject a dilute solution of the sample onto the column. For a given mass of sample, it has been found more advisable to inject a larger volume of a dilute solution onto the column to avoid localized overloading of the support [100, 169, 176]. As a rule of thumb, a sample volume of up to one quarter to one third of that of the eluted peak of interest can be injected. It has also been found advantageous when injecting large volumes to use a solvent of such a polarity that the sample remains as a tight band at the top of the column before being eluted down the column with the mobile phase [181]. However, care must be taken when injecting the sample in a solvent other than the mobile phase that the sample is sufficiently soluble in the mobile phase to prevent precipitation at the top of the column [177, 181]. Only fully porous materials are recommended for preparative HPLC, as pellicular supports have low capacity. In a paper dealing with the factors which affect the loading capacity of preparative columns, Scott and Kucera pointed out that increased loading can be achieved by increasing the k' value of the solute and by increasing the surface area of the support [176]. As a reversed phase material was shown to have higher capacity than untreated silicas [174] reversed phase chromatography may be an even more efficient means of carrying out preparative HPLC, if the expense of bonded materials can be reduced.

In preparative HPLC, detector overload (when the linear dynamic range of the detector is exceeded) may be encountered resulting in truncated peaks [177, 181]. The effect can usually be avoided by switching to a different wavelength [174], but the problem is not serious as the recorded peak shape does not represent a loss in column performance.

The final step in a preparative experiment is to recover the sample from a large volume of solvent. If the eluent is a volatile organic solvent, the sample is easily obtained by evaporation of the eluent. If the eluent is aqueous, the sample will probably have to be extracted into a water-immiscible organic solvent, and the latter dried and evaporated to yield the solute. Due to the large volumes involved, solvent purity is thus important in preparative HPLC. Bristow has recently described a preparative chromatograph where the eluate collection is computer controlled [182]. In general, isocratic elution is preferred for preparative HPLC, although linear and step gradients [181] and flow programming [100, 101] have been successfully employed. Solvent recovery can be problematic when gradient systems are employed.

Preparative HPLC is one of the areas of HPLC where advances may be expected in the next few years, but already some interesting applications papers have appeared. Examples include the purification of cholesteryl phenylacetate (1 g per injection) on two coupled 500 x 23 mm i.d. columns of Spherosil

XOA-400 using a step gradient with hexane/dichloromethane/methanol eluents [181]. The flow rate was 30 ml min^{-1} and the pressure drop 1000 psi (6.8 MPa). Flow programming was used to separate N-allylaniline and N,N-diallyl aniline (100 mg total per injection) on a 500×10.9 mm i.d. column of Porasil A [100]. Preparative scale purifications of linoleyl acetate [183], cholesteryl phenylacetate [183] and progesterone [181] have also been reported and milligram amounts of the six major components of a pyrethrum extract were isolated [181]. Recently, a 10 g sample of a tetrahydrophenanthrene mixture was chromatographed [184]. The scaling-up procedure from analytical to preparative columns has been reported [183] and preparative high performance exclusion separations have also been published [101, 185, 186].

4.10   Operating tips

4.10.1   Column standardization and stability

When running a column with a range of complex samples and with a variety of different eluents some of which may have a harmful effect on the support (this applies especially to bonded phase materials) it is advisable to obtain a chromatogram of a test mixture with a simple mobile phase. After a series of runs with buffers, detergents etc. the column performance can then be checked with the simple test mixture. In this way, the irreversible adsorption of, for example, proteins from biological samples, or the stripping of the bonded phase can be diagnosed. As silica itself begins to dissolve at around pH 8 it is unwise to use either silica or bonded silica materials above this pH; chemically bonded phases also begin to be cleaved off around pH 2. With bonded phase materials oxidizing agents should also be avoided. The adsorption of polar or polymeric materials on silica will cause a gradual deterioration in performance. In some cases a precolumn may be the best method of protecting the analytical column from deterioration. Adsorbed compounds can usually be removed from silica columns by flushing with methanol or water and from bonded phase columns by flushing with methanol or dichloromethane.

4.10.2   Loss of efficiency or poor efficiency

When faced with a loss in efficiency check that the column packing material has not settled. If necessary top up to the correct level with a thick slurry of fresh material, tamp down and level off. Columns with irremovable top frits are rather unsatisfactory. Check also the injection system: make sure that the syringe needle injects at the correct position, or that the valve injector outlet tube is correctly positioned. Between the column bottom and the detector use the minimum length of 0.25 mm i.d. (or 0.125 mm i.d.) tubing to prevent peak dispersion and use zero dead volume ('drilled out') couplings.

Pre-tailed peaks usually indicate a channel in the column packing material which allows part of the solute to reach the column walls. This fraction travels down the column faster than the bulk of the solute, giving rise to pre-tailed peaks (or fronting). The usual remedy is to repack the column. In normal phase ion pair partition chromatography pre-tailed peaks also arise as a result of dissociation of the ion pairs in the organic phase [187].

Tailed peaks can be caused by using a valve injector which is not well swept out by eluent. Similarly, with syringe injectors, dead volume at the top of the column must be minimized for optimum peak shapes. Tailed peaks can also result from adsorption of polar compounds onto polar sites on the packing material (including some reversed phase materials). Mass overload and volume overload effects can also give rise to poor peak shapes [176].

4.10.3   The mobile phase

(a)   Pretreatment

The mobile phase should be filtered before being placed in the reservoir, and line filters should be used to prevent blockage of the valves in the pump. The mobile phase should also be freshly degassed before use (e.g. by refluxing) to remove dissolved air, otherwise gas bubbles can form in the detector cell.

Removal of contaminants from HPLC eluents may also be necessary to achieve UV transparency or satisfactory reproducibility. Common clean-up procedures include passing hexane down a column of dry silica gel to remove aromatics, passing ethers down a column of dry alumina to remove peroxides, and distillation to free solvents from antioxidants, plasticizers and polymeric impurities. Solvent purity is particularly important in preparative HPLC where the sample is usually recovered by evaporation of the solvent. For reproducibility in adsorption chromatography, the water content of the mobile phase may have to be carefully controlled as discussed later (p. 43).

(b)   Restrictions

Since most parts of the chromatograph in contact with the mobile phase are constructed of stainless steel or teflon, there are few restrictions on the nature of the mobile phase. However, halide ions should be avoided as these attack stainless steel especially at low pH values. Algal growth may result from the prolonged use of biological buffer solutions (e.g. citrate) and these usually have to be discarded. When working with buffer solutions and other potentially corrosive eluents, it is advantageous for both the chromatograph and the column packing material if the system is flushed out with, for example, water or a water/methanol mixture when not in use.

4.10.4   UV photometers

Bubble formation in the flow cell (usually 10 μl) of a UV photometer can be troublesome. Bubble formation is characterized by a very high absorbance or a wildly fluctuating absorbance reading, and degassing of solvents helps prevent this problem. Slight back pressure applied to the outlet of the flow cell will also reduce bubble formation in the cell. This can be achieved by using a length of narrow bore (0.25 mm i.d.) tubing, or by partially restricting the outlet tubing. When changing over columns, it is advantageous to flush the air bubbles out of the new column before connecting the latter to the detector.

A constant high or increasingly high absorbance reading may indicate contaminated solvents, dirty cell windows or an imminent lamp failure; or the lamp or flow cell may be misaligned.

Baseline drift can be caused by thermal effects and can be reduced by lagging or thermostatting the column, column outlet and detector.

4.10.5 Injection systems

Care should be taken to remove bubbles from the injection device especially in quantitative analysis. Care should also be taken to obtain a representative sample and any suspended material should be filtered off using a membrane filter. Syringes and injection valves should be thoroughly flushed between injections to prevent contamination of succeeding samples. For longer life, septa should be pierced before use and the syringe needle should be blunt-ended (or domed).

Blocked syringe needles are usually unblocked by threading a fine wire through the needle. Placing the blocked needle in an ultrasonic bath is often effective, or alternatively the blocked tip can be placed in a hot flame.

When disconnecting the injection device from the column after operation, ensure that pressure has fallen completely to zero otherwise the packed bed may be disturbed. (This applies also when disconnecting the column after slurry packing.)

4.10.6 Supports

With some '5 μm' materials it may be necessary to fractionate off 'fines' ($d_p < 3$ μm) to prevent blockage of frits or gauzes. For larger size ranges, the frits or gauzes used to retain the support should be of a sufficiently fine pore size or weave that they do not become blocked. Spherical particles tend to give columns of higher permeability (i.e. better flow properties) than irregularly-shaped materials [18, 128].

5

Modes of chromatography

Chromatographic separation of two components depends on their having differing k' values which in turn depends on their having differing distribution ratios between the stationary and mobile phases. The variety of stationary phases used in liquid chromatography (active adsorbent surfaces, polymer loaded glass beads etc.) results in a variety of separation modes. Some practical aspects of these modes are discussed below.

5.1   Adsorption chromatography

Adsorption chromatography is useful for separating non polar or fairly polar organic molecules. During the chromatographic run, the solutes are eluted down an adsorbent bed (usually silica or alumina) with an organic mobile phase (e.g. hexane, dichloromethane) and are retained according to their affinity for the adsorbent surface. Much of our understanding of the adsorption process is due to the pioneering work of Snyder, and a detailed acount of adsorbents and the mechanism of adsorption is given in his book [7]. Several other accounts of the nature of the adsorbent surface [188–191], the mechanism of adsorption chromatography [192–195] and practical details of adsorption chromatography [172, 196–199] have been given.

Silica, SiO_2, is a 3-dimensional lattice of SiO_4 tetrahedra, and at the silica surface the lattice is terminated by surface hydroxyl (silanol) groups. These silanol groups are slightly acidic and are responsible for the interactions of the silica surface with the solutes. There are about four to five silanol groups per nm^2 ($100Å^2$) [7, 188, 189]. The silica surface has a range of different types of silanol groups each with different affinities for solutes. Thus, even in an oversimplified model, free silanol groups, hydrogen-bonded silanol groups and geminal silanols (two hydroxyl groups on the same silicon atom) can be distinguished.

The surface of alumina is more complicated than that of silica [7], but it similarly has polar sites onto which solute molecules may adsorb. In adsorption chromatography, the solute molecules and the molecules of the mobile phase compete for the adsorbent sites. More polar molecules are more strongly

Table 5.1 *Solvent properties of chromatographic interest*

Solvent	$\epsilon°_{Al_2O_3}$ [a]	Viscosity* (mPa s, 20° C)	Density	Refractive index	Boiling point °C	UV cutoff nm
perfluoroalkanes	−0.25	0.39	1.70	1.267	~57.0	210
1,1,2-trichlorotrifluoroethane		0.69	1.517	1.357	47.6	230
pentane	0.00	0.23	0.629	1.358	36.2	210
2,2,4-trimethylpentane	0.01	0.50	0.692	1.404	99.3	210
hexane	0.01	0.33	0.659	1.375	69.0	200
cyclohexane	0.04	1.00	0.779	1.427	81.4	210
cyclopentane	0.05	0.47	0.740	1.406	49.3	210
1-pentene	0.08	0.24	0.640	1.371	30.0	215
carbon disulphide	0.15	0.37	1.260	1.626	46.2	380
carbon tetrachloride	0.18	0.97	1.590	1.466	76.8	265
n-pentyl chloride	0.26	0.43		1.413		225
m-xylene	0.26	0.62	0.864	~1.500	139.1	290
di-*n*-propyl ether[c]	0.28	0.37	0.747	1.368	89.64	220
2-chloropropane	0.29	0.33	0.862	1.378	35.74	225
toluene	0.29	0.59		1.496	110.6	285
1-chlorobutane	~0.30	~0.35	0.873	1.397	68.3	225
chlorobenzene	0.30	0.80	1.106	1.525	131.7	
benzene	0.32	0.65	0.879	1.501	80.1	280
bromoethane	0.37		1.460	1.424	38.35	
diisopropyl ether[c]		0.329	0.724	1.368	68.3	
diethyl ether[c]	0.38	0.23	0.713	1.353	34.6	220
diethyl sulphide	0.38	0.45	0.836	1.442	92.1	290
chloroform	0.40	0.57	1.500	1.443	61.2	245
dichloromethane	0.42	0.44	1.336	1.424	40.1	235
tetrahydrofuran[c]	0.45	0.55	0.880	1.408	66.0	215
1,2-dichloroethane	0.49	0.79	1.250	1.445	83.0	225
methylethylketone	0.51	0.40	0.805	1.381	79.6	330
1-nitropropane	0.53	0.84	1.001	1.400	131.2	380
acetone	0.56	0.32	0.818	1.359	56.5	330
dioxan	0.56	1.54	1.033	1.422	101.3	220
ethyl acetate	0.58	0.45	0.901	1.370	77.15	260
methyl acetate	0.60	0.37	0.927	1.362	57.1	260
1-pentanol	0.61	4.10	0.815	1.410	137.8	210
dimethyl sulphoxide	0.62	2.24	1.100	1.478	189.0	
aniline	0.62	4.40	1.022	1.586	184.4	
diethylamine	0.63	0.38	0.702	1.387	55.5	275
triethylamine		0.38	0.728	1.401	89.5	
nitromethane	0.64	0.67	1.138	1.394	101.2	380
acetonitrile	0.65	0.37	0.782	1.344	82.0	190
pyridine	0.71	0.94	0.983	1.510	115.3	305
dimethoxyethane		0.45	0.863	1.380	83.0	220
o-dichlorobenzene[d]		1.324	1.306	1.552	180.48	
1,1,1-trichloroethylene[e]		0.566	1.460	1.476	87.19	
N,*N*-dimethylformamide[e]		0.92	0.949	1.428	153.0	
tetrachloroethylene[e]		0.90	1.620	1.505	121.0	
N-methylpyrrolidone[e]		1.65	0.819	1.470	202.0	
1-methylnaphthalene[d]			1.025	1.618	235.0	
dimethylacetamide			0.937	1.438	165.0	

Table 5.1 (continued)

Solvent	$\epsilon^{\circ}_{Al_2O_3}$ [a]	Viscosity* (mPa s, 20° C)	Density	Refractive index	Boiling point °C	UV cutoff[b] nm
trans-decahydronaphthalene (decalin)			0.896	1.470	194.6	
trifluoroethanol[e]		~1.20	1.390	1.219	73.6	
2-methoxy ethanol	0.74	1.72	0.965	1.401	124.6	220
isopropanol	0.82	2.30	0.786	1.380	82.3	210
ethanol	0.88	1.20	0.789	1.361	78.5	210
methanol	0.95	0.60	0.796	1.329	64.7	205
1,2-ethanediol	1.11	19.90	1.114	1.427	197.3	210
acetic acid	Large	1.26	1.049	1.372	117.9	
water	Large	1.00	1.000	1.330	100.0	

[a]Data taken from reference 7. Adsorption energies (solvent strengths) on silica ($\epsilon^{\circ}_{SiO_2}$) follow the same order as those on alumina: individual values are approximately 0.77 times the alumina value.
[b]Wavelength at which transmission falls to 10% for good commercial grade solvents, glass distilled.
[c]Gives rise to problems with peroxides.
[d]Eluent for exclusion chromatography (at elevated temperatures).
[e]Eluent for exclusion chromatography.

Table reproduced with permission from Bristow, P. A. (1976), Liquid Chromatography in Practice, H.E.T.P., Macclesfield. U.K.

*Viscosity conversion is 1 mPa s = 1 cP = 1 mN s m^{-2}

adsorbed and the use of increasingly more polar eluents will tend to elute the solutes more quickly from an adsorbent column. Snyder has arranged the solvents according to their adsorption energies (ϵ^0 values) on alumina [7]. This list, shown in Table 5.1, is known as the eluotropic series and has important practical utility in adsorption chromatography. Separation between two otherwise unresolved solutes can often be achieved by holding the solvent strength (ϵ^0 value) constant while varying the solvent composition, the increase in selectivity being due to secondary solvent effects. For example, the ϵ^0 values of the binary mixtures pentane/pyridine (95:5), pentane/acetone (74:26) and pentane/dichloromethane (23:77) are equal ($\epsilon^0_{alumina} = 0.40$) [7]. The largest secondary solvent effects are obtained for binary mixtures whose components have the greatest difference in ϵ^0 values. Thus of the three mixtures quoted, the pentane/ pyridine mixture would give the largest secondary solvent effects.

In adsorption chromatography, traces of water in the mobile phase can be extracted out onto the adsorbent surface. Since solute molecules will not now be adsorbed on these sites the surface is deactivated. Problems of reproducibility of retention times can be encountered if say 'dry' hexane is used in conjunction with an active (i.e. essentially water-free) column of silica or alumina, since traces of water will adsorb on the surface of the support and lead to shortening of the retention times [175, 200]. For this reason, and to reduce peak tailing, it is sometimes preferable to add a small amount of a more polar component to the

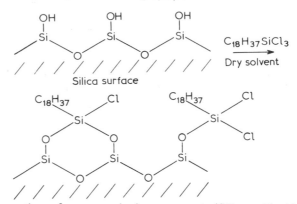

Fig. 5.1 Preparation of a reversed phase support. (Silicon-chloride bonds remaining at the end of the reaction are rapidly hydrolyzed to silanol groups on contact with moisture.)

organic mobile phase (e.g. 1 per cent methanol in hexane); alternatively, 50 per cent water-saturated hexane can be used. This is prepared by mixing equal volumes of dry hexane and water-saturated hexane. The hexane is dried by passing it down a column of silica previously heated at typically 150° C in a vacuum oven. Water-saturated hexane is prepared by passing hexane down a column of silica previously conditioned with 25 per cent wt/wt of water. Scott and Kucera have developed a series of twelve solvents of polarities increasing from heptane to water for use with silica for the chromatography of complex mixtures [201]. Using this incremental gradient elution technique, a k' range of 10^4 relative to heptane was covered.

Adsorption chromatography has been widely used for the chromatography of polycyclic aromatic compounds, phenols and amines. Since silica is acidic, amines tend to be very strongly adsorbed and give tailed peaks; accordingly for basic compounds alumina is usually preferred. Conversely, acidic compounds such as phenols chromatograph well on silica but not so well on alumina. Problems arise if the molecules to be separated are highly polar or contain ionizable groups, in which cases poor peak shapes are almost inevitable. The problem may be reduced somewhat, e.g. in the chromatography of polar acids on silica, by adding an acidic component to the mobile phase to suppress the ionization of the solutes and block the more active silanol groups, but in general reversed phase chromatography offers a more useful solution.

5.2   Liquid—liquid partition chromatography

In liquid—liquid partition chromatography a chromatographic support is coated with a layer (e.g. 1 per cent by wt) of a polymer such as polyethylene glycol (PEG) or a liquid which is insoluble in the mobile phase, e.g. β, β'-oxydipropionitrile (BOP) which is used with hexane as mobile phase. The relative solubility of the solutes in the mobile and stationary phases determines the k' values and efficient separations have been reported with this mode of chromatography

[9, 202—206]. A disadvantage of liquid—liquid partition chromatography is that a precolumn is necessary to saturate the mobile phase with stationary phase to avoid stripping the stationary phase, and this requires careful thermostatting of the column and precolumn. Also, gradient elution is not possible with liquid—liquid partition systems, although flow programming has been used [102]. It is probable that chromatography on bonded phases will usually be an attractive alternative to liquid—liquid partition chromatography.

5.3   Reversed phase chromatography

As mentioned previously (p. 44) molecules which are highly polar give rise to problems of long retention times and peak tailing in adsorption chromatography. The solution to this problem lies in chromatographing the compounds using a non-polar stationary phase in conjunction with a polar mobile phase, a technique introduced by Howard and Martin [207]. For example, diamond [208] and hydrocarbon polymers coated on pellicular materials [209] have both been used as non polar supports. The most common method of performing these separations has been by using a hydrocarbon-bonded surface in conjunction with a polar eluent such as a methanol/water mixture. The polar molecules now have little affinity for the hydrophobic support and are eluted relatively quickly by the aqueous mobile phase. Since the most polar molecules will now tend to elute first, this type of chromatography is referred to as reversed phase chromatography. The most common bonded hydrocarbon phase is that formed by bonding octadecylsilyl groups ($C_{18}H_{37}Si-$) groups to silica, as illustrated in Fig. 5.1, although shorter chain lengths are commercially available. The efficiencies obtained with reversed phase materials can be similar to those obtained with the unbonded silica adsorbent (Fig. 2.4). Examples of a normal phase separation and a reversed phase separation of the same components are shown in Figs. 5.2 and 5.3. The mechanism of reversed phase chromatography has not been satisfactorily worked out, although several papers have appeared on the subject [136, 150, 152, 160, 208, 210—213]. One of the problems is that most commercially available materials may contain a high proportion of residual silanol groups [174] and retention on such supports will almost certainly be by a mixed adsorption/partition mechanism. (The residual silanol groups may be silanol groups originally present on the silica surface which have not reacted for steric or other reasons, or they may be formed on a di- or trichlorosilyl reagent at the end of the reaction by hydrolysis of unreacted Si—Cl bonds, as shown in Fig. 5.1.) The calculation of the surface coverage of a bonded phase is illustrated in Appendix 6. It seems reasonable to suggest that blocking of residual silanol groups with, for example, trimethylsilyl groups should reduce the tailing of polar solutes on these sites in reversed phase chromatography. It has been suggested that the role of the bonded hydrocarbon is to extract from a binary mobile phase the more lipophilic component [150]. This results in an organic-rich layer at the particle surface where the chromatographically useful partitioning takes place. (The unexpectedly high capacity of an ODS-bonded phase relative to a

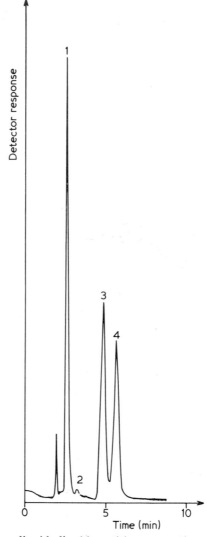

Fig. 5.2   Normal phase liquid–liquid partition separation of a steroid mixture. Column packing, Zipax +1% BOP; column dimensions, 1 m x 2.1 mm i.d.; eluent, heptane; pressure 600 psi (4MPa); flow rate, 1 ml min^{-1}; column temperature, ambient; detection, UV at 254 nm. Solutes: 1, progesterone; 2, androsterone; 3, testosterone; 4, 19-nortestosterone. Reproduced with permission from reference 107.

range of silica gels may be further evidence of this effect [174].) This feature of chromatography on bonded phases also accounts for the versatility of reversed phase materials. For example, if ODS-silica is used with a methanol/water eluent containing a small amount of a long chain alkyl cyanide the cyanide will be preferentially partitioned into the hydrophobic ODS surface and the material would act as a dynamic cyano phase. Similar considerations apply in reversed phase ion pair partition chromatography and soap chromatography (Section 5.5) where organic ions added to the eluent may be partitioned into the organic surface to form dynamic ion exchangers.

Regarding the chain length of reversed phase materials, the longer the chain length, the greater the solute retention with a given eluent. Selectivity has been found to increase with chain length [151], to remain constant [154] and also to show an initial increase (C_{12} chain compared with a C_6 chain) followed by a levelling off (comparing C_{12} and C_{18} chains) [144]. In practice, many separations obtained on long chain materials can be reproduced on short chain materials by increasing the amount of the aqueous component in the eluent; in addition, short chain materials have proved useful for the separation of highly polar and ionic solutes in the reversed phase and ion pair partition modes [150]. The use of carbonaceous supports for reversed phase chromatography may provide some interesting materials for future separations [214].

The use of reversed phase HPLC in correlating the chemical structures of drugs and pesticides with their biological activity is discussed in Appendix 7.

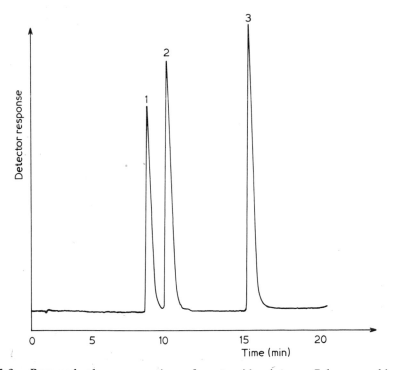

Fig. 5.3 Reversed phase separation of a steroid mixture. Column packing, Permaphase ODS; column dimensions, 1 m x 2.1 mm i.d.; eluent, linear gradient from water to water/methanol (50:50) at 5 per cent min^{-1}; pressure, 1000 psi (6.8MPa), flow rate, 1 ml min^{-1}; column temperature, 50°C; detection, UV at 254 nm. Solutes: 1, 19-nortestosterone: 2, testosterone; 3, progesterone. Reproduced with permission from reference 107.

5.4 Ion exchange chromatography

In ion exchange chromatography, the chromatographic support contains ions which are capable of being exchanged with ionic solutes in the mobile phase. Ion exchangers in HPLC normally utilize a bonded quaternary ammonium group $[-(CH_2)_n\,NR_3^+X^-]$ for the separation of anions, and a bonded sulphonic acid group $(-C_6H_4SO_3^-H^+)$ for the separation of cations. The technique of ion exchange chromatography was utilized over 25 years ago for the separation of nucleic acid components [215], amino acids [216] and sugars [217]. Comprehensive texts on the theory [218] and applications [219] of ion exchange chromatography have appeared, and the uses of ion exchange chromatography in HPLC have been reviewed [220, 221].

In ion exchange chromatography there are two important variables which the chromatographer can alter to achieve satisfactory separation between two or more components. Firstly, by increasing the ionic strength (μ) of the eluent (a buffer solution), the solutes will elute faster if an ion exchange mechanism is operating. In fact $k'\propto 1/\mu$ and experimental plots of the reciprocal of the ionic strength vs capacity ratio have given linear plots [150]. This is useful in deciding the optimum ionic strength for a given separation.

Secondly, variation in the pH of the buffer should cause a linear change in log k' if the acid or base is weakly ionized. This tends to be true only over limited pH ranges in practice [150] and depends on the pK values of the acids or bases being chromatographed, as well as on the degree of ionization of the ion exchange groups. In complex problems, trial runs of the individual components at various values of pH and ionic strength of the buffer solution is a quick route to the understanding of the behaviour of the system.

Although some efficient analyses of amino acids have been reported using polymeric ion exchange resins [6, 222] (in spite of the use of the post-column ninhydrin reaction detector) poor mass transfer can be a problem with such resins. For this reason, chemically bonded microparticulate ion exchangers are the most efficient ion exchange materials currently available. However, the efficiencies obtained in ion exchange chromatography have not equalled the efficiencies obtained with adsorbents or reversed phase materials, and ion pair partition chromatography is fast becoming the preferred method of chromatographing ionic solutes.

5.5 Ion pair partition chromatography

Ion pair partition chromatography is an alternative method to ion exchange chromatography for the separation of ionic or ionizable molecules, and credit for the development of the technique is due largely to Schill and co-workers [187, 223]. Three fairly distinct types of ion pair partition chromatography have evolved and these are described in turn.

5.5.1 Normal phase ion pair partition chromatography
In normal phase ion pair partition chromatography silica is used as the support

and is coated with an aqueous stationary phase containing an acid such as perchloric acid (for the separation of bases) or a base such as tetraethyl-ammonium hydroxide (for the separation of acids); an organic mobile phase is used, a typical example being butanol/chloroform. In the separation of, say, organic bases, R_3N, undissociated ion pairs $(R_3NHClO_4)_{org}$ predominate in the organic mobile phase while discrete ions exist in the aqueous stationary phase according to

$$(R_3NHClO_4)_{org} \rightleftharpoons R_3NH_{aq}^+ + ClO_{4\ aq}^- \qquad (5.1.)$$

In practice, normal phase ion pair partition chromatography is rather troublesome due to stripping of the stationary phase and the necessity of a pre-column loaded with the aqueous phase. (The advantage of using a UV absorbing counter ion in ion pair chromatography has already been mentioned in Section 4.3.) A more convenient alternative is reversed phase ion pair partition chromatography.

5..5.2   Reversed phase ion pair partition chromatography

In reversed phase ion pair partition chromatography a chemically bonded silica support (e.g. ODS-silica) is used along with an aqueous phase containing the required acidic or basic counter ions. Thus it is an easy matter to change the nature or concentration of the counter ions. This overcomes the problems of reproducibly coating the support with aqueous phase and the associated problems of bleeding. Ion pairs are now partitioned into the stationary phase and the free ions eluted by the mobile phase.

5..5.3   Soap chromatography

Soap chromatography is the term given to the form of ion pair partition chromatography where the counter ion of the ionizable components to be separated is a detergent [224]. For example, cetrimide $(C_{16}H_{33}N^+Me_3)$ has been used for acids and lauryl sulphonic acid $(C_{12}H_{25}SO_3^-)$ for bases. High efficiencies have been obtained using reversed phase ion pair partition chromatography and soap chromatography for many ionic compounds such as sulphonic acid dyestuffs, which would be otherwise difficult to chromatograph [224]. To develop a reversed phase ion pair partition (or soap) separation in practice, the pH of the mobile phase is adjusted such that the solute(s) are in the ionized form. The ratio of the organic component of the mobile phase (e.g. methanol or acetonitrile) to the aqueous component is adjusted so that the solute(s) are virtually unretained. The addition of a small amount of ion pairing agent (e.g. 0.01–0.1 per cent) should cause a dramatic increase in solute retention. The separation can then be optimized by variation in the concentration and nature of the organic component of the eluent, the pH, ionic strength and nature of the aqueous component, the operating temperature, and the nature and concentration of the counter ion. A particular advantage of reversed phase ion pair partition and soap chromatography is that since aqueous eluents are used, biological samples can be injected with minimal clean-up. In soap (and reversed phase ion pair partition) chromatography, extraction of the counter ion pair (e.g. $C_{12}H_{25}SO_3Na$) into the hydrophobic support surface means that an ion pair exchange mechanism (5.2) rather than an ion pair partition mechanism (5.1) commonly operates.

$$[C_{12}H_{25}SO_3^-Na^+]_{org} + R_3NH_{aq}^+ \rightleftharpoons [C_{12}H_{25}SO_3^-R_3NH^+]_{org} + Na^+. \quad (5.2)$$

As more applications are published demonstrating the high efficiency of the various types of ion pair partition chromatography it seems likely that this mode of chromatography will be preferred over ion exchange chromatography for many separations involving ionizable compounds.

5.6 Exclusion chromatography

(Liquid) exclusion chromatography is the preferred name [225a,b] given to separations which occur due to differences in the molecular weights (and shapes) of the solutes being separated. Historically the term gel permeation chromatography was used to describe such separations occurring in organic solution e.g. tetrahydrofuran with a polystyrene matrix, and gel filtration was used to describe separations carried out in aqueous solution in conjunction with, for example, dextran or agarose gels. Various reviews on exclusion chromatography have appeared [226–229].

For exclusion chromatography to take place, the pore size of the support (agarose, dextran, polystyrene beads, silica or glass), should be of such a range as to allow some of the solute molecules to enter. Very small molecules will permeate all the internal pores of the support particle; very large molecules will be unable to enter any of the pores (i.e. these will be totally excluded). Thus as a polymer sample of mixed molecular weights flows down the column there will be a separation of the solutes according to molecular weight. The high molecular weight, totally excluded solute will elute first, and the lower molecular weight species which are held back as a consequence of permeating the particle pores, will elute in reverse order of molecular weight. The only conditions required of the solvent in exclusion chromatography are that it should completely dissolve the solute at the operating temperature of the column and that it should 'wet' the particle surface. The pore volume of the particles should be as large as possible, to increase the separation power of the support. It is also necessary that there is no adsorption of the solutes onto the particle surface (e.g. proteins on silica) otherwise the separation mechanism will not be one of exclusion and erroneous results may be obtained. Steps may be taken to reduce adsorption e.g. by silanization of the silica surface [230, 231] or by using a mobile phase of similar constitution to the particle matrix (e.g. a toluene/polystyrene combination). Since the most retained solute in exclusion chromatography is a small molecule which permeates the particle completely (e.g. a solvent molecule), this means that even in a complex sample all the components are eluted in a relatively short (and predictable) time, in contrast to the other modes of HPLC.

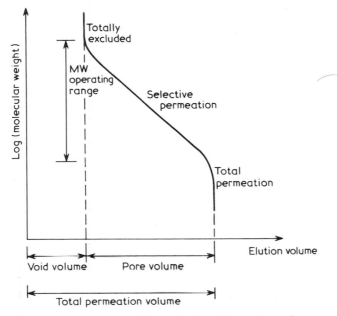

Fig. 5.4　Idealized calibration curve for exclusion chromatography.

This however also means that in exclusion chromatography the peak capacity (the number of resolved peaks in the chromatogram) is rather limited.

In order to use exclusion chromatography for the assignment of molecular weights in polymer samples, a calibration curve is constructed, ideally using molecules of a similar chemical composition to those being investigated. An idealized calibration curve is shown in Fig. 5.4. The useful range for size separations lies between the 'totally excluded' and the 'totally permeating' volumes, and in this region the plot of log (molecular weight) *vs* elution volume should be linear if molecular weight assignments are to be correct. If the slope of the plot is steep in this region, there will be poor resolution between similar molecular weight fractions, but a wide molecular weight range will be covered; if the slope is shallow, good resolution will be obtained over a rather narrow range of molecular weight.

An experimental calibration curve obtained by Attebery [186] is shown in Fig. 5.5 and demonstrates the importance of molecular shape as well as molecular weight. Thus, reserpine with an actual molecular weight of 608 has an apparent molecular weight from the graph of 410 due to its compact structure which makes it appear smaller in solution. Vitamin A palmitate with its rod-like structure has a larger apparent molecular weight (760) than expected (543). To overcome this effect of molecular shape, a calibration curve of log (hydro-dynamic volume) *vs* elution volume, can be constructed which takes account of the variations in the shapes of the solutes. (The hydrodynamic volume is the product of the molecular weight and the intrinsic viscosity of the solute.)

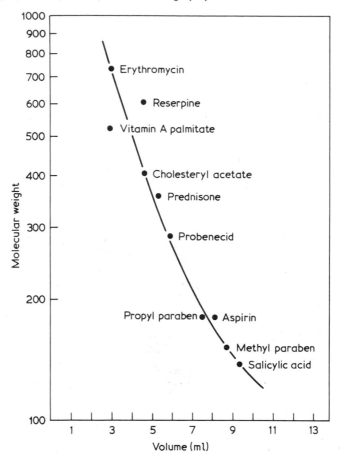

Fig. 5.5 Experimental calibration curve for exclusion chromatography. Column packing, 100 Å μStyragel (10 μm); column dimensions, three coupled 300 x 7 mm i.d. columns; eluent, tetrahydrofuran; flow rate, 2 ml min^{-1}. Reproduced with permission from reference 186.

Materials for exclusion chromatography have included mechanically unstable gels such as dextrans and agarose which swell in some solvents and do not withstand pressure. Polystyrene beads (styrene cross-linked with divinylbenzene) are available for use at moderate pressures, but it seems that future advances will centre on rigid packings based on silica or porous glass. It is to be expected that new materials of the latter type will become available with high pore volumes, useful ranges of pore size and their attendant mechanical stability. Such materials will allow size separations to be performed more efficiently than is at present possible. Some commercially available materials for exclusion chromatography are mentioned in Appendix 8.

5.7 Affinity chromatography

The technique of affinity chromatography is really outside the scope of this text, but due to its similarity to the other modes of HPLC (especially to ion exchange chromatography) and its enormous biological potential, brief mention will be made here. In affinity chromatography, a ligand which binds specifically to one particular compound or a small group of compounds is bonded to an insoluble support such as polystyrene beads or silica. When a complex biological mixture is passed down a column of such a material, only these compounds with an affinity for the bonded group are retained. After elution of the other components of the mixture, the retained compounds of interest can be eluted by a change in mobile phase. The technique has been used to purify enzymes, proteins, viruses and nucleic acids. Reviews on the biological [232] and chemical [233] aspects of affinity chromatography have appeared.

Part II
The application of HPLC in pharmaceutical analysis

6

Drug analysis by HPLC

6.1 Introduction

6.1.1 Comparison of HPLC with other methods of drug analysis

Several distinct problems occur in the analysis of drugs. These may entail interference from excipients during the assay of raw materials and formulated products; difficulties in the determination of small amounts of degradation products in a vast excess of the parent drug; and finally, and frequently the most difficult, the analysis of drugs and their metabolites in body fluids. In the last case nanogram amounts of drug may have to be assayed in the presence of a wide variety of endogenous compounds.

The commonly used methods of quantitative drug analysis utilize spectroscopic measurements (UV/visible, fluorescence), chromatographic assays (GLC, HPLC and TLC), microbiological procedures and radioimmunoassay techniques. Disadvantages of gas chromatographic assays of drugs include the elevated operating temperatures required which may occasionally cause thermal degradation of the underivatized drug, and the frequent necessity for derivatization to increase volatility and to improve chromatographic behaviour.

The gas chromatographic analysis of drugs and their metabolites in body fluids frequently necessitates considerable sample clean-up (normally solvent extraction) to prevent interference from endogenous substances. Although such difficulties may be frequently minimized by the use of specific detection systems, some degree of pre-analytical purification is almost invariably required. Mass spectrometric detection affords very high specificity which may enable still further reduction in sample preparation, but such systems require a very large capital outlay along with a substantial maintenance budget, and a skilled operator. Gas chromatographic methods are generally inapplicable for the analysis of highly polar substances (e.g. sulphate and glucuronide conjugates of drugs and metabolites) or of substances of high molecular weight.

Thin layer chromatographic methods are inexpensive but lack resolution and are difficult to quantitate, although using multiple spotting techniques several assays can be performed simultaneously. Recently, high performance TLC systems have been developed which may overcome some of the traditional disadvantages of this technique [234].

Spectroscopic (UV/visible) methods for drug analysis are based on the measurement of absorbance at a characteristic wavelength. The method is fairly non-specific and will not usually discriminate between closely related metabolites or degradation products. In the assay of formulated products, the drug may have to be extracted from UV/visible absorbing excipients prior to spectroscopic analysis. In the analysis of body fluids the presence of large amounts of UV/visible absorbing endogenous components necessitates lengthy extraction procedures which may not always remove all the interference. Similar limitations apply to fluorescence methods.

Microbiological assays are time consuming and often lack the ability to discriminate between closely related molecules such as the various tetracyclines. Radioimmunoassay methods, although relatively easy to perform, are difficult to develop and although capable of detecting very low levels of substrate, $(10^{-9} - 10^{-12} \text{g ml}^{-1})$, they frequently lack specificity. In addition each assay measures only one component of a mixture.

HPLC offers many advantages in drug analysis but also has a few drawbacks. One important advantage is that extraction procedures and sample clean-up prior to injection are greatly reduced relative to most methods. Thus in the analysis of drugs in formulated products it is frequently sufficient to crush or mix the product with a solvent, filter and inject [235–237]. Excipients move with the solvent front or are separated chromatographically from the drug of interest. A considerable advantage of the technique is the versatility in the choice of mobile phase. Thus in the analysis of a drug in urine injected directly onto the column, a reversed phase system may be chosen such that the aqueous mobile phase will elute most of the UV absorbing endogenous components with the solvent front. Manipulation of the other components of the mobile phase will then cause the drug to elute in a transparent section of the chromatogram. Several examples are discussed in which untreated urine or plasma samples are injected onto the column without marked deterioration in column performance [238–240]. The capability of HPLC to handle directly polar drugs and their metabolites and conjugates in body fluids is another major advantage of the method. Rapid sample clean-up by passage down an exclusion column to achieve a molecular weight separation may also be useful. Since analyses can be performed at room temperature, thermal degradation is extremely rare. Also, the high resolving power of the system means that individual members of a compound class can be discriminated. Thus toxic metabolites or by-products can be analyzed. Automated injection systems are available for HPLC systems so that in quality control situations the method, once established, is not operator intensive. Analyses are usually completed in 10–30 min. Quantitation of components is usually achieved by the inclusion of a suitable internal standard in the sample. However, in some instances the reproducibility of injection using a valve injector has been found to be adequate for quantitative analysis without an internal standard [241, 242]. Preparative HPLC has also shown promise for the collection of the separated components in amounts sufficient for identification.

The disadvantages of HPLC centre around the detection systems available. Ultraviolet spectrometers are the most commonly used detectors but require that the compound has a UV absorbing chromophore. Variable wavelength UV spectrophotometers offer reasonable versatility but some steroids and other drugs must be derivatized before UV detection is possible.

Another slight disadvantage may be that the chemically bonded stationary phases which are most applicable to drug analysis should only be used within the pH range 3–7 to guarantee long term stability.

In the last few years HPLC has indeed become a widely used technique in the field of drug analysis. Applications in pharmaceutical analysis [243, 244], clinical chemistry [245], and forensic science [246] have been reviewed. Several applications of HPLC to pharmaceutical analysis are discussed in the following sections, including drug stability studies [242, 247, 248], the determination of trace impurities or decomposition products in bulk drug samples [237, 249, 250] and cross-contamination studies [251]. Many examples are also discussed of the assay of drugs and metabolites in body fluids. The use of reversed phase HPLC in structure/activity correlations of drugs is discussed in Appendix 7.

6.1.2 Chromatographic systems for drug analysis

Some of the points made earlier about column packing materials are summarized for convenience. A wide variety of microparticulate column packing materials in the optimum size range for practical use (~5 μm) are commercially available. These materials produce columns of high efficiency, e.g. at least 2000 theoretical plates for a 100 mm ion exchange column and up to 8000 theoretical plates for a 100 mm column of silica, alumina, a reversed phase material (e.g. an ODS-silica) or an ion pair partition column. One problem often encountered in the analysis of polar drugs by adsorption chromatography is poor, badly tailed peaks. The addition to the mobile phase of small quantities of an amine such as triethylamine in the chromatography of basic compounds, or an acid such as acetic acid in the chromatography of acids, often leads to improved peak shapes. This is due partly to suppression of ionization of the solute and partly to a deactivation of the most strongly adsorbing sites of the support, producing a correspondingly more homogeneous adsorbent surface. One advantage of adsorption chromatography is that a TLC method can be more readily translated into an HPLC one than with reversed phase or ion exchange systems.

Injection of body fluids onto columns of adsorbents is not recommended since protein and other polar compounds adsorbed by the top of the column result in a gradual deterioration of column performance. (A removable pre-column may help overcome this deficiency.) Reversed phase columns, reversed phase ion pair (or soap chromatographic) systems and ion exchange systems are less affected by injection of body fluids and are more suited to such analyses. In particular, the two former systems give very high efficiencies, require little attention in use and are recommended. Efficiencies in ion exchange systems, even with the modern 5 μm chemically bonded materials, have not yet approached those of adsorption or reversed phase systems.

Liquid–liquid partition chromatography has been used for some elegant separations but in practice requires more attention due to problems in maintaining the equilibria between the mobile and stationary phases. Accordingly this mode of HPLC seems less useful in routine situations.

Larger particle sizes of adsorbents and chemically bonded stationary phases can be used with only a modest loss of efficiency. Use of the polymer coated glass bead type of ion exchanger or reversed phase material results in a further loss in efficiency and the use of polystyrene/divinylbenzene ion exchange resins in modern HPLC should be avoided because of the poor efficiencies associated with the bad mass transfer characteristics of these materials. As an example of the increase in efficiency possible over the past few years, two separations of

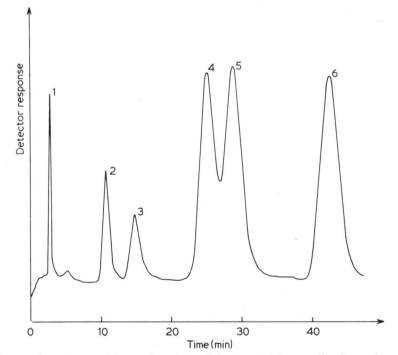

Fig. 6.1 Separation of barbiturates. Column packing, pellicular anion exchanger; column dimensions, 3 m x 1 mm i.d.; eluent, linear gradient from 0.1 mM to 1.0 mM sodium chloride, pH 7; pressure, 700–900 psi (4.8–6.1 MPa); flow rate, 0.43 ml min^{-1}; column temperature, 80°C; detection, UV at 254 nm Solutes: 1, ketohexobarbitone (0.3 μg); hydroxyamylobarbitone (13 μg); 3, contaminant; 4, amylobarbitone (5.3 μg); 5, phenobarbitone (2.5 μg); 6, hydroxyphenobarbitone (3.5 μg). Reproduced with permission from reference 252.

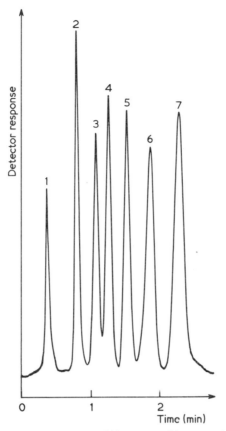

Fig. 6.2   Separation of barbiturates. Column packing, methyl silica (5 μm); column dimensions, 100 x 2.8mm i.d.; eluent, methanol/water (50:50); column temperature, ambient; detection, UV at 205 nm. Solutes: 1, potassium chromate; 2, phenobarbitone; 3, barotal; 4, butobarbitone; 5, hexobarbitone; 6, vinbarbitone; 7, quinalbarbitone. Reproduced with permission from reference 253.

barbiturates are shown in Figs 6.1 and 6.2 In Fig. 6.1 a polystyrene-based ion exchange resin was used at a column temperature of 80°C [252]. In Fig. 6.2 a modern reversed phase material was used at room temperature [253]. Thus it should be borne in mind that many of the separations discussed in the following sections will be capable of marked improvement.

6.2   Antibiotics

The quantitative determination of antibiotics is one of the more difficult areas of pharmaceutical analysis. The most serious disadvantage of many of the microbiological, spectrophotometric and chemical methods in current use is their lack of specificity. Gas chromatographic methods are fast and specific but the technique is only applicable to those antibiotics which are thermally stable after derivatization. Several assays based on HPLC have been reported in the last few years. Since many antibiotics contain ionizable groups, the earliest separations were obtained by ion exchange chromatography [251, 254, 255] but more recently reversed phase methods on bonded phase materials have been used [255–260].

6.2.1 Penicillins and cephalosporins

(a) Standard mixtures and pharmaceutical preparations
Penicillins G and V (Fig. 6.3) and some synthetic penicillins such as ampicillin, oxacillin, methicillin and nafcillin have been analyzed by reversed phase chromatography on C_{18}/Porasil B using 0.05M ammonium carbonate/methanol (70:30)* as mobile phase [256]. Because of the relatively large particle size (37–75 μm) of the packing, rather broad peaks were obtained, but adequate separations were achieved. Ampicillin, penicillin G and penicillin V have also been separated on a 1 m long column of Vydac P150 AX ion exchange resin with a mobile phase of 0.02M sodium nitrate in 0.01M pH 9.15 borate buffer [242]. UV detection at 254 nm was used throughout. Since small quantities of penicillins can cause allergic reactions in sensitive individuals it is useful in cross-contamination studies to be able to analyze traces of penicillins in other drugs. Ampicillin has been analyzed as a contaminant in nitrofurantoin preparations using anion exchange chromatography with an eluent of 0.01M potassium dihydrogen phosphate buffer at pH 6.5 [251]. The limit of detection of this procedure was 1.0 μg.

Several commercially available cephalosporin antibiotics (Fig. 6.3) have been separated on an ODS-SIL-X-II column with an eluent of 0.95M ammonium carbonate/methanol (95:5) [256]. The method was used for the quantitative analysis of cephradine in an oral suspension, using cephaloglycin as an internal standard. The average recovery (four replicate samples) was 100.0 per cent and the coefficient of variation was 1.09 per cent.

A preparative separation of a 500 mg crude sample of cefazolin has also been achieved on a 1.2 m x 8 mm i.d. column of C_{18}/Porasil B with an aqueous eluent [256]. Sufficient quantities of three impurities were obtained for identification by other techniques.

(b) Drug stability studies
The analysis of degradation products in commercial penicillins has twofold importance; firstly, in pharmacokinetic studies it is desirable to distinguish between the drug and any degradation products and secondly, allergic reactions attributed to penicillin may frequently be caused by such compounds. Accordingly it is essential to be able to detect the presence of these compounds in the pharmaceutical preparation.

*Mobile phase components are given as a volume/volume ratio unless otherwise stated.

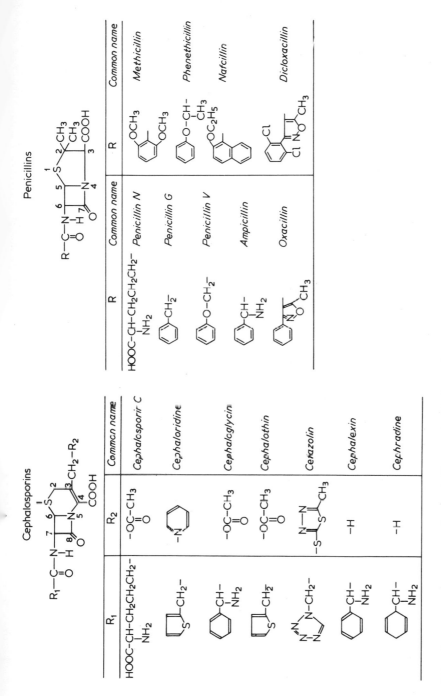

Fig. 6.3 Structure of representative penicillins and cephalosporins.

The quantitative analysis of potassium penicillin G in the presence of five of its degradation products, DL-penicillamine, benzylpenilloic acid (two isomers), benzylpenamaldic acid, benzylpenicilloic acid and benzylpenillic acid has been achieved on a 600 x 2.3 mm i.d. column of Bondapak AX/Corasil with a pH 3.8 buffer [247]. The peaks were detected at 254 nm. Using this procedure it was possible to detect these substances in aged solutions of potassium penicillin G.

Ampicillin may be differentiated from its degradation products by anion exchange chromatography on Vydac P150 AX with a borate buffer at pH 9.15 [242]. The analytical purity of ampicillin trihydrate bulk powders and other pharmaceutical products containing ampicillin, calculated using this method, showed no significant difference from those obtained by the iodimetric method. After ageing of powders at 55°C for one month the HPLC assay showed 6.8–12.5 per cent reductions in potencies. No such loss was detected by the iodimetric method, but a microbiological assay agreed very well with the HPLC results.

(c) Metabolism studies
The high degree of resistance of cefoxitin to biodegradation by bacterial cephalosporinases has been attributed to the presence of the 7 α-methoxyl group [261]. In order to study the bioavailability of both cefoxitin and cephalothin it was necessary to develop an analytical method capable of detecting and determining the amount of deacylated metabolite in the presence of the unchanged drug in urine. Cephalothin and deacetylcephalothin have been analyzed on an anion exchange resin at elevated temperatures [262]. Subsequently a method was developed for the analysis of both drugs along with their deacylated metabolites which involved chromatography at room temperature, and reduced the possibility of on-column degradation [241]. Healthy humans were dosed with cefoxitin or cephalothin and a series of urine samples was collected before and at intervals after dosing. The samples were filtered and injected directly onto a 1 m x 3 mm column of Zipax SAX Permaphase. The mobile phase was 0.25M sodium acetate at pH 5.0 and detection was performed at 254 nm. Peak heights were used for quantitation, and the recovery was calculated as the percentage of the administered dose in the urine. The analysis of 86 urine samples for intact cefoxitin or cephalothin by both the HPLC and bioassay methods showed good agreement, indicating that there was no significant interference from natural urinary constituents or drug metabolites. The HPLC method permitted simultaneous determination of drugs and metabolites and indicated that both drugs were excreted rapidly after direct intravenous infusion. Substantial excretion of deacetylcephalothin occurred during the first hour after dosing, whereas cefoxitin was only slightly metabolized to decarbamylcefoxitin.

6.2.2 Polypeptide antibiotics
In an initial study of the analysis of the structurally similar polypeptides in the antibiotic bacitracin, more than 22 components were separated in 40 min on a

1 m column of Bondapak C_{18}/Corasil using gradient elution [263]. Only the peaks corresponding to bacitracin A, B and C were found to have antimicrobial activity.

In an improved method, baseline resolution of the polypeptides was achieved using a 300 x 4.6 mm i.d. stainless steel column of μBondapak C_{18} with a convex gradient programme [257]. The antimicrobial potencies calculated using the HPLC assay agreed well with those obtained by the microbiological method. Bacitracin powders stored at room temperature for 6—7 years showed a fall in potency of around 50 per cent as analyzed by both HPLC and the microbiological methods [263]. The effects of processing methods to reduce microbial contamination levels were also evaluated [257]. Thus it was found that ^{60}Co irradiation at 1.8 Mrad resulted in no loss in potency or change of any bacitracin components, whereas heating caused a 35 per cent loss of anti-microbial activity and a two- to eight-fold increase in bacitracin F, the oxidative degradation product. Ethylene oxide gas treatment also caused considerable loss of potency (46 per cent) but in this case the chromatograms showed no significant increase in bacitracin F.

Peptide antibiotics of the polymyxin group have also been analyzed on a μBondapak C_{18} column using linear gradient elution [257]. UV detection at 254 nm was used to monitor polymyxins B_1 and B_2 and a moving wire FID for circulin A and B and colistin A and B (i.e. polymyxin E_1 and E_2).

Actinomycin complexes are chromopeptides differing only in limited areas of their peptide moieties. Several actinomycins have been chromatographed on a 1.8 m x 2.3 mm i.d. column of Bondapak C_{18}/Corasil with acetonitrile/water (50:50) as mobile phase [264]. Actinomycins C_1, C_2 and C_3 were completely resolved in 35 min. In the presence of cis-4-chloro-L-proline, the organism S. parvullus synthesizes actinomycins containing one or both of the proline residues replace by the chloroderivative. The three major components produced by this organism have been analyzed under the same conditions.

These methods are of particular significance since few HPLC procedures are so far available for the analysis of polypeptides. Further examples are discussed in Chapter 12.

6.2.3 Polyene antifungal antibiotics

Polyene antifungals show little activity against bacteria in contrast to the antibacterial macrolide antibiotics to which they are chemically related. Polyenes are classed as tetraenes, pentaenes, hexaenes or heptaenes, depending on the number of conjugated double bonds present in the chromophore. Group separation is relatively easy but differentiation of members of each class is significantly more difficult, and various techniques (paper chromatography, TLC, counter current distribution, GLC and chemical degradation) have only been partially successful in resolving this problem. The analysis of the polyenes is made more difficult by their poor solubility in water and organic solvents, and their sensitivity to heat, air oxidation, UV and visible light, and pH values outside the range 6—9.

(a)   Tetraenes
The tetraene antifungals are subdivided into large and small macrolactone ring antibiotics, the most important commercially being nystatin, a large ring macrolactone. Nystatin can be separated into its components by chromatography on Vydac RP with an aqueous methanol/tetrahydrofuran mobile phase [265]. The separation took 15 min in contrast to the 12—14 h required to resolve the components by counter current distribution.

(b)   Pentaenes
Of the pentaenes, eurocidin, which from earlier mass spectral data was known to exist as a mixture of two components, may be completely resolved by reversed phase chromatography in 5 min [265]. Injection of 25 ng of the mixture gave easily identifiable peaks at 350 nm.

(c)   Heptaenes
Candidin can be separated into five components in 14 min on Vydac RP whereas only three components had been obtained previously after extensive counter current distribution separation [265]. Amphotericin B shown in Fig. 6.4 is the clinically most important and chemically best defined non-aromatic heptaene antibiotic. It was found to consist of one main component containing two trace impurities [265]. The three aromatic heptaene antibiotics, candicidin, levorin and trichomycin have been resolved into their individual components on a 1 m x 2.1 mm i.d. column of Permaphase ODS (20—37 μm) using gradient elution at 55°C [266]. Candicidin and levorin appeared to be identical, each containing the same mixture of seven components. Trichomycin was also a complex mixture, but different from the other two.

The detector of choice for the analysis of polyenes is a variable wavelength UV spectrophotometer. The wavelengths of maximum absorption occur at regular intervals from 291—361 nm for the tetraenes through to the heptaenes.

6.2.4   Macrolide antibiotics
The liquid chromatographic analysis of macrolide antibiotics has been examined [267]. The eight components of a leucomycin complex from *S. kitasatoensis*

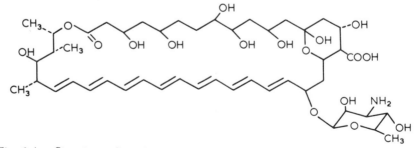

Fig. 6.4   Structure of amphotericin B.

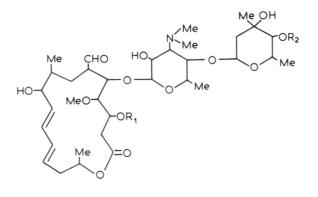

	R$_1$	R$_2$
Leucomycin A$_1$	H	COCH$_2$CH(CH$_3$)$_2$
A$_3$	COCH$_3$	COCH$_2$CH(CH$_3$)$_2$
A$_4$	COCH$_3$	COCH$_2$CH$_2$CH$_3$
A$_5$	H	COCH$_2$CH$_2$CH$_3$
A$_6$	COCH$_3$	COCH$_2$CH$_3$
A$_7$	H	COCH$_2$CH$_3$
A$_8$	COCH$_3$	COCH$_3$
A$_9$	H	COCH$_3$

Fig. 6.5   Structure of the components of a leucomycin complex.

shown in Fig. 6.5 have been resolved by reversed phase chromatography. The samples were detected at 243 nm. The same system has also been used to separate the components of spiramycin, tylosin, magnamycin and erythromycin [267].

6.2.5   *Tetracyclines*
The tetracyclines are an extremely important group of chemotherapeutic agents with antibacterial activity towards a broad range of pathogenic microorganisms. They have proved to be a difficult group of molecules to analyze by HPLC, and indeed by other methods. Problems in the HPLC analysis of the tetracyclines arise from the large number of polar functional groups which the molecules possess. These functional groups tend to adsorb strongly on the silanol sites of support materials (including the residual silanol sites of many chemically bonded and polymer coated supports), and gave rise to tailed, badly shaped peaks. Furthermore, tetracycline itself is unstable in solution and epimerizes to the 4-epi form. It has also been necessary in some studies to wash stainless steel columns and fittings with ethylenediaminetetraacetic acid (EDTA) to prevent complexing of the tetracycline with the metal. The preparation of silyl

derivatives prior to analysis by GLC requires carefully controlled reaction conditions if the tetracyclines are not to be decomposed by the silylating reagents, and the microbiological assay is of inadequate specificity. Tsuji investigated the reversed phase analysis of tetracyclines and found Zipax ODS and Corasil C_{18} to be unsatisfactory [268]. However, he did obtain separations of tetracycline, anhydrotetracycline, 4-epitetracycline, 4-epianhydrotetracycline, chlortetracycline and doxycycline using a Zipax HCP column, although peak shapes were poor, and the column and packing material had to be conditioned with EDTA [268]. Similarly, separation of oxytetracycline, doxycycline, demeclocycline and tetracycline on ODS-SIL-X-II required preconditioning of the column with EDTA [256]. An improved method for the analysis of tetracycline was obtained by chromatography on a 300 x 4.6 mm column of ODS-silica with an acetonitrile/0.02M phosphate buffer (pH 2.5) gradient system [260]. The analysis of a tetracycline sample took 16 min compared to 25 min with the previous systems, and there was no need to condition the column with EDTA. The recovery of tetracycline from pharmaceutical products was better than 99.6 per cent and the coefficient of variation for the analysis of tetracycline powders was 0.66 per cent.

In a systematic study of support materials for tetracycline analysis Knox and Jurand confirmed that reversed phase materials held the most promise for these separations, preliminary results being superior to those obtained with adsorbents, ion exchange materials and liquid—liquid partition systems [258]. They obtained the best separation of five tetracyclines, namely tetracycline, 4-epi-tetracycline, chlortetracycline, epi-anhydrotetracycline and anhydrotetracycline on a microparticulate silica whose surface silanol sites had been fully silanized. The eluent consisted of perchloric acid/acetonitrile and it was considered that the acetonitrile was extracted from the mobile phase by the hydrophobic surface of the support. The tetracyclines were assumed to partition as their perchlorate ion pairs into this organic layer. The system was used for the quantitative analysis of 10 μg of tetracycline containing anhydrotetracycline (2 per cent), epi-anhydrotetracycline (0.6 per cent), chlortetracycline (2 per cent) and epi-tetracycline (5 per cent) as impurities.

An automated procedure for the determination of tetracyclines has also been reported [269]. Twelve samples per hour could be analyzed on a single column.

The analysis of rolitetracycline (2-N-pyrrolidinomethyltetracycline) and its hydrolysis product tetracycline in rolitetracycline and rolitetracycline nitrate formulations was carried out on a 2.25 m x 1.8 mm i.d. column of Pellionex CP-128 pellicular cation exchange resin with UV detection at 254 nm [270]. An aqueous ethanol buffer at pH 4.35, containing EDTA, was used as eluent. Analyses were performed on formulations containing ascorbic acid, magnesium gluconate, or lidocaine HCl. The latter two components were not resolved from tetracycline but since they had only a low UV absorption, a suitable correction could be applied when they were present.

A method has been developed for the determination of anhydrotetracycline

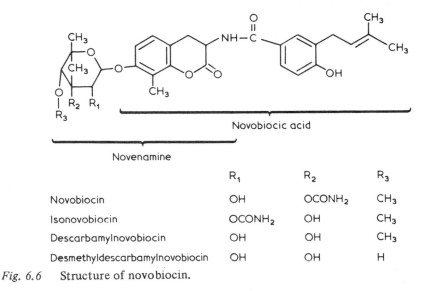

	R$_1$	R$_2$	R$_3$
Novobiocin	OH	OCONH$_2$	CH$_3$
Isonovobiocin	OCONH$_2$	OH	CH$_3$
Descarbamylnovobiocin	OH	OH	CH$_3$
Desmethyldescarbamylnovobiocin	OH	OH	H

Fig. 6.6   Structure of novobiocin.

and epi-anhydrotetracycline on a cation exchange resin using a neutral 0.3*M* EDTA buffer as eluent [271]. The estimated limits of detection at 429 nm were 12 ng of anhydrotetracycline and 7 ng of epi-anhydrotetracycline.

β-Cetotetrine has been determined in biological fluids using reversed phase HPLC with electrochemical detection [272]. Chromatography was performed on a 600 x 2 mm i.d. column of Bondapak Phenyl/Corasil (37–50 μm) using an eluent of dipotassium hydrogen phosphate (0.025*M*), sodium EDTA (0.1 per cent) buffer, pH 7.8/ethanol (93:7). The limit of detection was approximately 100 pg injected on-column.

As yet no completely satisfactory method has been developed for the efficient routine analysis of these very important drug substances.

6.2.6   *Miscellaneous*

(a)   *Novobiocin*
The antibiotic novobiocin and some related components shown in (Fig. 6.6) have been analyzed on a 1 m column of Zipax HCP with a mobile phase of 0.02*M* phosphate buffer at pH 7.0/methanol (85:15) [273]. Sodium novobiocin was found to contain on average 91.4 per cent novobiocin, 1.8 per cent isonovobiocin, 1.5 per cent dihydronovobiocin, 4.6 per cent descarbamylnovobiocin, 0.1 per cent novobiocic acid, 0.5 per cent novenamine and <0.1 per cent desmethyldescarbamylnovobiocin. The HPLC method was sensitive to 30 ng of novobiocin, and the coefficient of variation for the determination was 0.51 per cent (*n* = 7). The potency of 13 novobiocin powders analyzed by the HPLC method was compared with the potencies calculated from the microbiological

assay, and no statistically significant differences were observed. Novobiocin may, of course, be analyzed with good precision by GLC. However, it is less specific than HPLC since an amide or a glycosidic bond is cleaved during the derivatization procedure, rendering it impossible to differentiate between isonovobiocin and novobiocin.

(b) Kanamycin

Kanamycins A and B are aminoglycoside antibiotics produced by fermentation. Although the fermentation of each compound is selective there may be a necessity to monitor one in the presence of the other. Since they are difficult to differentiate by chemical or biological methods, an analytical method based on HPLC has been developed. The kanamycins have been separated on a 1 m x 2 mm column of Zipax SCX (37–44 μm) using a mobile phase of 0.01M ammonium phosphate at pH 9.1 [274]. A fluorescence detector was used after post-column on-line derivatization with fluorescamine or o-phthalaldehyde. Using this method less than 0.35 μg of kanamycin B could be detected in as much as 7 μg of kanamycin A injected. The analysis time was less than 15 min per sample.

(c) Daunomycin

Quantitative determination of the plasma levels of daunomycin and its main metabolite, daunomycinol, the structures of which are shown in Fig. 6.7, has been achieved by HPLC using adriamycin as an internal standard [275]. The separations were obtained on a silica microsphere column using a quaternary solvent mixture consisting of dichloromethane/methanol/water/25 per cent ammonia (90:9:0.8:0.1). The components in the eluted peaks were detected at 490 nm and no interfering peaks were observed. The method was sensitive to plasma levels of daunomycin and daunomycinol above 19 ng ml^{-1}.

(d) Penimocycline

Penimocycline is an antibiotic obtained by Mannich reaction between tetracycline and ampicillin. The separation of penimocycline from tetracyclines and other impurities was most effectively achieved by liquid–liquid partition chromatography on a MicroPak CH column using gradient elution with 0.02M phosphate buffer at pH 7.6 containing 1 mM EDTA [259].

(e) Chloramphenicol

Reversed phase chromatography of chloramphenicol, a widely used antibiotic, has been reported [256] on ODS-SIL-X-II using 0.5M ammonium carbonate as the mobile phase. C_{18}/Corasil and Vydac RP could also be used as stationary phases but Permaphases ODS and ETH and Phenyl/Corasil were found to give insufficient retention. The purity of chloramphenicol is greatly dependent on the purity of intermediate products in its preparation. Three of the by-products formed in the chloramphenicol production process were initially separated on a

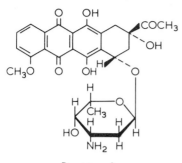 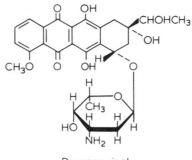

Daunomycin   Daunomycinol

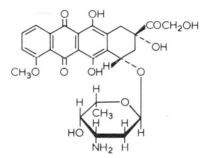

Adriamycin

Fig. 6.7   Structure of daunomycin, daunomycinol (its main metabolite) and adriamycin.

1 m column of Zipax SCX using 0.05M sodium sulphate (pH 1.1) as eluent [254]. Improved methods were subsequently developed whereby six inter-mediates could be separated on MicroPak-CN [255] and MicroPak-NH₂ columns [276].

(f)   Griseofulvin
The antifungal agent griseofulvin is produced by fermentation. A crude fermentation mixture contains many closely related compounds and is usually difficult to analyze. Griseofulvin and four other structurally related components from such a mixture have been separated on a column of Permaphase ETH using hexane/chloroform (95:5) [235]. Griseofulvin has also been analyzed on a column of ODS-SIL-X-II [256].

A specific and quantitative liquid–solid chromatographic method has been developed for the determination of 6-demethylgriseofulvin, a metabolite of griseofulvin in urine [277]. The method involved extraction of the sample, after hydrolysis of the glucuronide, into 20 per cent *n*-butanol in benzene, addition of the internal standard, 4-(*m*-hydroxyphenyl)-1-isopropyl-7-methyl-2(1H)-quinazoline, and analysis on a 314 x 2 mm i.d. column of Pellosil using a heptane/alcohol gradient system. Detection was effected at 254 nm, and a sensitivity limit of 6 μg ml^{-1} was obtained. Griseofulvin, if present, could be determined simultaneously.

6.3 Antibacterials

6.3.1 Sulphonamides

The sulphonamides constitute a very important class of antibacterial agents which are still widely prescribed, although for many diseases their use has been superseded by antibiotics. Traditionally, these drugs were analyzed by paper chromatography and colourimetry [278], but faster and simpler methods have been developed with the advent of HPLC.

Sulphonamides have been separated by ion exchange chromatography on both anion [279] and cation exchange resins [280], ion pair partition chromatography [224, 281–283] and adsorption chromatography [284] on silica. These methods have been applied to the analysis of the drugs in pharmaceutical dosage forms [280, 285, 286], body fluids [287, 288] and tissues [59].

(a) Standard mixtures

Twenty-one sulpha drugs were analyzed on a 1 m x 2 mm i.d. column of Zipax SAX using an eluent of 0.01M sodium borate (pH 9.2) containing sodium nitrate at various concentrations [279]. Examination of the effect of the ionic strength of the mobile phase upon retention times and peak widths resulted in a scheme that could be applied to the separation of many combinations of sulphonamides.

Adsorption systems have also been found useful. The retention data of nineteen sulphonamides on a 250 x 4 mm i.d. column of Spherisorb S5W (5 μm) have been reported [284]. The mobile phase was cyclohexane/anhydrous ethanol/glacial acetic acid (85.7:11.4:2.9). Acetic acid was added to prevent tailing of the peaks. Separations were performed at ambient temperature and the eluent was monitored at 260 nm.

Ion pair partition chromatography may also facilitate the efficient separation of sulphonamides [224, 281–283]. Tetrabutylammonium was a suitable counter ion with n-butanol/heptane as the mobile phase [282]. Wide variations in k' and α were possible by changing the mobile phase composition, counter ion composition, pH or ionic strength. Silica columns coated with buffered aqueous solutions of tetrabutylammonium sulphate resulted in efficiencies of 4000–6000 theoretical plates per 250 mm and were stable for long periods of time. The separation of 13 sulphonamides using this system is shown in Fig. 6.8. Sulphonamides can also be separated by reversed phase ion pair chromatography using pentanol or butyronitrile coated on a support of LiChrosorb RP-2 (10 μm), and a mobile phase of 0.01M aqueous tetrabutylammonium hydroxide (pH 7.9) [224]. The separation of sulphadiazine, sulphalene, sulphamerazine, sulphamethoxazole and sulphasomidine on a butyronitrile stationary phase has been achieved in 15 min. Using this system sulphadiazine and sulphamethoxasole could also be separated from their N^4-acetyl derivatives.

(b)   Assays of sulphonamides in pharmaceutical preparations
The separation and quantitative analysis of sulphamerazine, sulphadiazine and sulphadimidine in pharmaceutical dosage forms present a difficult problem to the pharmaceutical analyst. Paper chromatographic and colourimetric methods are slow and tedious. Poet and Pu [280] have reported the determination of these sulphonamides by HPLC on a 1 m × 6.35 mm column of Zipax SCX cation exchanger. The mobile phase was 0.2M disodium phosphate solution adjusted to pH 6.0 with 85 per cent phosphoric acid. At a flow rate of 0.7—0.8 ml min^{-1} the sample components were separated in 15—20 min. Using sulphadimethoxine as an internal standard for quantitation, analytical data were obtained for four representative lots of tablet formulation and two suspension formulations. The calculated coefficients of variation for replicate sample injections ranged from 0.9 to 4.0 per cent.

The analysis of sulphasalazine in bulk powder and tablet form has been achieved by reversed phase chromatography on a 1.2 m × 300 mm i.d. column of

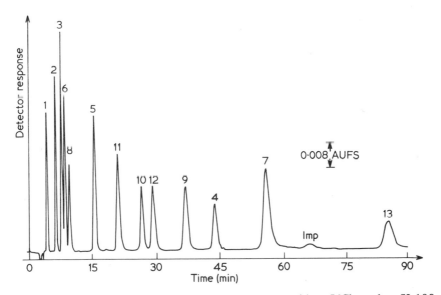

Fig. 6.8   Separation of sulphonamides. Column packing, LiChrospher SI 100 (10 μm) loaded with 0.3M TBA HSO$_4$ in 0.1M phosphate buffer, pH 6.8; column dimensions, 250 × 3.2 mm i.d.; eluent, heptane/n-butanol (75:25); flow rate, 0.4 ml min^{-1}; column temperature, 27 ± 0.5°C; detection, UV at 254 nm. Solutes: 1, phthalyl sulphathiazole; 2, sulphabenzamide; 3, sulphisoxazole; 4, sulphacetamide; 5, sulphadimethoxine; 6, sulphachloropyridazine; 7, sulphadiazine; 8, sulphaquinoxaline; 9, sulphamerazine; 10, sulphamethoxypyridazine; 11, sulphathiazole; 12 sulphamethazine; 13 sulphapyridine. Reproduced with permission from reference 282.

C_{18}/Corasil using a mobile phase of phosphate buffer (pH 7.7)/isopropanol (9:1) [285] The assay method utilized a simple one-step solubilization procedure with dimethylformamide, addition of internal standard (propylparaben) and chromatography with detection at 254 nm. The method was specific for sulphasalazine in the presence of its precursors or degradation products.

Sulphacetamide and its principal hydrolysis product, sulphanilamide, may be determined in ophthalmic solutions with sulphabenzamide as an internal standard [286]. Chromatography was performed on a column of Corasil II (37–50 μm) using an eluent of dichloromethane/isopropanol/conc. ammonia (26:13:0.5).

(c) Assays for sulphonamides in body fluids and tissues

A reversed phase chromatographic method has been applied to the determination of sulphadimidine residues in bovine kidney, liver, muscle and fat tissue at the level of 0.04 μg g^{-1} [289]. Chromatography was performed on a column of Bondapak C_{18}/Porasil B using 2.5 per cent isopropyl alcohol in phosphate buffer (pH 7.7) as mobile phase.

A similar method has been used to screen the urine of cattle treated with sulphadimidine, sulphamerazine and sulphathiazole [287]. Standard samples of the three drugs and their N^4-acetylated metabolites could be completely resolved on an amino bonded phase column using a methanol mobile phase. Analysis of the compounds in urine extracts, however, was apparently less successful but nevertheless permitted identification and quantitation. Using UV detection at 254 nm, injections containing 0.2 μg of each individual sulphonamide or their metabolites could be quantitated.

6.3.2 Non-sulphonamide antibacterial and antiseptic agents

(a) Assays of pharmaceuticals

An automated method for the determination of chlorhexidine (Hibitane) and its salts in formulated pharmaceutical products has been described [290]. Methanolic extracts were chromatographed on a 100 x 4 mm i.d. column of Partisil (11 μm) at 25° C with acetonitrile/0.02N sulphuric acid (91.5:8.5) as eluent. The sample was automatically introduced onto the column by use of a slide valve. The procedure was applied to a range of production and development samples and gave results indistinguishable from the conventional colourimetric assay. The precision of the method was checked on a sample of liquid antiseptic and a coefficient of variation of 1.45 per cent (n =10) was obtained.

Sondack and Koch have analyzed the antibacterial agent cinoxacin (Fig. 6.9,I) in capsule and ampoule formulations [291]. Chromatography was performed on a 1 m x 2.1 mm i.d. column of Zipax SAX using a pH 9.2 buffer as eluent, and UV detection at 254 nm. The method was also applicable to the determination of nalidixic acid (Fig. 6.9,II) and oxolinic acid (Fig. 6.9,III) in

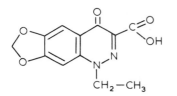

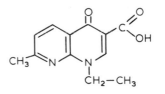

I

II

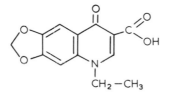

III

Fig. 6.9   Structure of cinoxacin and related antibacterial agents. I = cinoxacin, II = halidixic acid, III = oxolinic acid.

aluminium hydroxide gel suspensions. Sulphanilic acid may be used as the internal standard for the determination of cinoxacin and oxolinic acid and, for the estimation of nalidixic acid, sulphamerazine was used. Analysis of twenty replicate capsules of cinoxacin gave values which were on average 0.9 per cent higher than the label claim, with a coefficient of variation of ±2.0 per cent. Recoveries of 100 per cent were obtained from spiked placebo samples.

The components of an antitubercular mixture, isoniazid and pyridoxine HCl may be separated in 12 min by an ion pair chromatographic method [292]. The compounds were chromatographed on a column of μBondapak C_{18} using a mobile phase of methanol/water (60:40), (pH 2.5), containing 0.01M dioctyl sodium sulphosuccinate. Isonicotinic acid was used as the internal standard for quantitation and the drugs were detected at 293 nm.

(b)   Assays in body fluids
An HPLC approach has been used for the analysis of conjugates of 5-chloro-7-iodo-8-quinolinol (Cloquinol or Chinoform) in human urine [240]. Separations were effected on a 500 x 2 mm i.d. column of Zipax SAX using gradient elution and UV (254 nm) detection. Urine was directly injected onto the column and no interfering peaks were observed.

Flucytosine (5-fluorocytosine), an antifungal agent, may be determined in blood and urine by chromatography on a pellicular cation exchange resin at 78° C [293]. Ammonium phosphate buffer at pH 2.0 was used as eluent and the detection wavelength was 280 nm. Aliquots of serum and urine were examined directly using a precolumn to protect the analytical column from contamination. Ion exchange chromatography also permits the determination of nalidixic acid and its metabolite, hydroxynalidixic acid, in plasma and urine extracts [294]. The analysis was performed on a 500 x 2.1 mm i.d. column of Zipax SAX with 0.05M sodium sulphate/0.02M boric acid at pH 9 as eluent. The limit of detection of either compound in plasma extracts was around 2 ng injected on-column, corresponding to a plasma concentration of 0.25 μg ml^{-1}.

6.4 Antidepressants

6.4.1 Dibenzazepines and related compounds

Knox and Jurand have studied the separation of twenty common tricyclic psychosedative drugs, including nortriptyline, amitriptyline, protriptyline, opipramol, imipramine and trimipramine by ion pair partition and adsorption chromatography [295]. Ion pair chromatography was effected with a stationary phase of 0.1M perchloric acid/0.9M sodium perchlorate loaded in situ on LiChrosorb SI 100 (10 μm) and a mobile phase of n-butanol/chloroform (70:30). This system was found to be most effective in producing group separations. Adsorption chromatography on Spherisorb A20Y (20 μm) alumina with dichloromethane/hexane/acetic acid mobile phases of various compositions, was much more selective for individual members of a single group. Adsorption chromatography on silica gel has also been shown to be an effective method for resolving tricyclic antidepressants [296].

An HPLC method has been described which permits the identification of nortriptyline and amitriptyline at levels approximating to those found in human plasma during treatment [297]. The use of protriptyline as internal standard allowed accurate quantitative determination of the other two compounds. Recoveries after extraction from aqueous solution were 95—105 per cent in the presence of the internal standard, and calibration curves for both amitriptyline and nortriptyline were linear over the range 0.01—5.5 μg. Separations were effected on a 250 x 2 mm i.d. column of MicroPak SI 10 using methanol/ammonia (100:1.5) as the mobile phase. The eluate was monitored at 240 nm. The separation of trimipramine, chlorimipramine, imipramine and opipramol may also be obtained by adsorption chromatography on LiChrosorb SI 60 using a mobile phase of dichloromethane/isopropanol (90:10) saturated with aqueous ammonia [298]. Alternatively, chlorimipramine and desmethylchlorimipramine have been determined in plasma samples (using trimipramine as internal standard) by chromatography of the ion pairs formed with tetraethyl ammonium chloride [299]. The stationary phase, 25 per cent v/v of a mixture of 0.1M hydrochloric acid and 0.1M tetraethylammonium chloride, was coated on Dia Chrom (37—44 μm) and a mobile phase of hexane/isobutanol (9:1) at 0.5 ml min^{-1} was used. The coefficients of variation were around 4 per cent for chlorimipramine (n = 20; range 25—150 ng ml^{-1}) and 6 per cent for desmethyl-chlorimipramine (50—300 ng ml^{-1}).

6.4.2   Xanthines

Xanthines are important stimulants present in tea, coffee and cocoa and their analysis is mainly of interest to the food industry. Caffeine, theobromine and theophylline have been separated by partition chromatography on Corasil coated with 1.1 per cent Poly G-300 using heptane/ethanol (100:10) as mobile phase [300]. Resolution of these compounds plus xanthine was also achieved on a 1 m x 2 mm column of Zipax SAX using a mobile phase of 0.01M sodium borate [301].

(a)   Theophylline

Theophylline is an important therapeutic agent frequently used in the treatment of asthma. Clinical effects of the drug are related to its blood concentration with maximum therapeutic activity occurring in the $10-20 \mu g \, ml^{-1}$ range. Toxic symptoms are increasingly observed above this level and it is necessary to have a fast precise method of analysis to use in emergency determinations. Several rapid HPLC methods for the analysis of theophylline in body fluids have been reported.

One method utilized as little as 0.1 ml aliquots of serum which are prepared for analysis simply by diluting with a solution of the internal standard, 8-chlorotheophylline [239]. A precolumn was used to protect the analytical column from contamination. In another procedure, acidified plasma or serum samples were extracted with chloroform/isopropanol containing oxazepam as the internal standard [302]. Methods which involve extraction of theophylline are not ideal since the partition coefficient of theophylline between organic solvents and an aqueous phase is low. Alternatively plasma which has been filtered through a membrane to remove proteins may be directly injected on to the column [303].

Several chromatographic procedures have been used to separate theophylline from the dietary xanthines and uric acids. These include ion-exchange [239], adsorption [302, 304, 305] and reversed phase [303] chromatography. No interference from co-administered drugs are observed in any of the methods [239, 302–305]. However, it is noteworthy that heparin interferes during the determination of theophylline in plasma [239].

An HPLC method of studying the pharmacokinetics of theophylline in infants from whom only small volumes of blood are available has been developed [304]. Only 0.5 ml of plasma was required for each determination and a detection limit of around $0.1 \mu g \, ml^{-1}$ was obtained. The method involved chromatography on a 500 x 3 mm i.d. column of MicroPak SI 10 (10 μm) silica using chloroform/isopropanol/acetic acid (84:15:1) with detection at 273 nm. Good precision and linearity of response was observed down to $40 \mu g \, ml^{-1}$. The limit of detection for theophylline at 254 nm after chromatography on Aminex A-5 cation exchange resin with an ammonium phosphate buffer at pH 3.65 was around 20 ng [239]. Two other non-clinical reports of the separation of theophylline from related molecules have appeared [306, 307].

Analysis by HPLC is in general faster, requires less sample preparation and is considerably more specific than UV spectrophotometric analyses. Comparable results were obtained with assays based on HPLC, UV [302, 303] and GLC [303].

(b)   Uric acid

Kissinger and co-workers have described the estimation of uric acid in serum by HPLC using an electrochemical detector [308, 309]. Serum samples were chromatographed on anion exchange columns, with a 10 mm replaceable precolumn to trap proteins. The eluate was oxidized electrochemically and accurate measurements were possible on 100 pg of uric acid [308]. No interferences from endogenous material were noticed. Because of increased selectivity, the HPLC method using the electrochemical detector was found to be an improvement over the existing colourimetric and enzymic assays [309].

6.4.3   Amphetamines

Stimulants of this type have been considered in Section 6.8.

6.5   Central nervous system depressants

6.5.1   Hypnotics and sedatives

Some of the most commonly used sedatives are barbiturates. There is special interest in the rapid analysis of these drugs since they are frequently encountered in overdose cases. Analyses are normally performed by GLC but this necessitates the use of a derivatization step prior to examination of the extract. Several analytical methods for barbiturates involving HPLC have been reported [223, 252, 253, 310–312].

Since barbiturates are weak acids, early liquid chromatographic separations were obtained on ion exchange resins [252, 311]. Rapid separation of barbiturates may be achieved on a 1 m x 2.1 mm i.d. column of Zipax SAX utilizing alkaline (0.01M sodium borate/0.03M sodium nitrate) or acid (0.01M acid) mobile phases [311]. Superior results were obtained with the alkaline eluent, but both systems were satisfactory for qualitative or quantitative analysis of pharmaceutical preparations. The limit of detection using a UV monitor at 254 nm was around 1 ng.

Simpler, more efficient methods have been developed for barbiturate analysis with the improvement of column packings and packing techniques. Either normal phase [298] or reverse phase [223, 253] methods may be employed. The resolution of four barbiturates by a reversed phase ion pair partition method [223] is shown in Fig. 6.10.

A suitable method for the determination of trace amounts of barbiturates in blood and saliva has been developed [253]. Extracts were analyzed on a 100 x 2.8 mm i.d. column of methyl silica using methanol/water (50:50) as eluent and UV detection at 220 nm. Detection limits of around 3 ng were obtained.

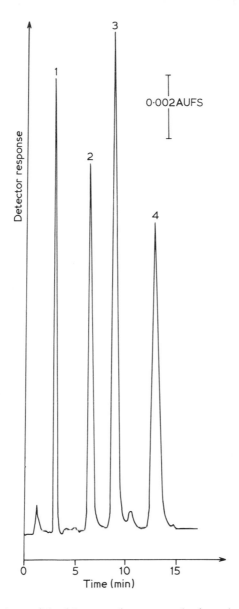

Fig. 6.10 Separation of barbiturates by reversed phase ion pair chromatography. Column packing, LiChrosorb RP-2 (10 μm) loaded with butyronitrile; column dimensions, 150 x 3.2 mm i.d.; eluent, 0.01M tetrabutylammonium (TBA), pH 7.7; pressure, 400 psi (2.7MPa); linear flow rate, 1.6 mm sec^{-1}; column temperature, ambient; detection, UV at 254 nm. Solutes: 1, barbitone; 2, allobarbitone; 3, aprobarbitone; 4, phenobarbitone. Reproduced with permission from reference 223.

The analysis of phenobarbitone along with other anticonvulsant drugs is considered in Section 6.5.2.

The non-barbiturate sedative, ethchlorvynol can be converted via an unsaturated aldehyde to a semicarbazone derivative which has a chromophore at 280 nm [313]. A chromatographic method for the determination of ethchlorvynol in rat serum makes use of this specific reaction for detection of the drug [314]. The derivative was chromatographed on a column of μBondapak C_{18} using aqueous alcohol mobile phases. The limit of detection by this procedure was 50 ng ml^{-1}. Several drugs, often taken in conjunction with ethchlorvynol, such as diazepam, some phenothiazines and barbiturates, did not interfere with this assay.

6.5.2 Anticonvulsants

Several methods for the determination of anticonvulsant drugs in body fluids have been described [310, 312, 315–317].

Simultaneous monitoring of several drugs such as phenobarbitone, phenytoin, primidone, ethosuximide, methsuximide and carbamazepine can be accomplished by reversed phase chromatography on a column of ODS-SIL-X-I with a mobile phase of water/acetonitrile (83:17) [317]. The eluted drugs were detected by their absorption at 195 nm giving detection limits of 0.1– 0.5 μg ml^{-1} for 0.5 ml serum samples. Procedures have also been described for the simultaneous determination of phenobarbitone and phenytoin on columns of microparticulate silica with detection at 254 nm [310, 312].

Carbamazepine is metabolized to its 10,11-epoxide which exerts about the same anticonvulsant activity as the parent drug. It is thus important to be able to measure both the parent drug and its epoxide metabolite when correlating the plasma levels of the drug and its effect. One suitable procedure for the determination of both the drug and its metabolite involved chromatography of a plasma extract on a column of Durapak Carbowax 400/Corasil with an eluent consisting of hexane/dichloromethane/dimethylsulphoxide (76:22.8:1.2) [315]. However, this procedure is complicated by the fact that other anticonvulsant drugs such as phenytoin, primidone and phenobarbitone interfere with the assay and must be removed by an alkaline wash prior to analysis. Using UV detection at 254 nm, 4 ng and 40 ng of carbamazepine and its metabolite respectively could be measured. Alternatively, as shown in Fig. 6.11, the compounds could be rapidly determined without interference from other anticonvulsants on a column of LiChrosorb SI 100 with a mobile phase of dichloromethane/ tetrahydrofuran (95:5) and detection at 250 nm [316]. A detection limit of about 2 ng ml^{-1} was obtained for carbamazepine using a 1 ml plasma sample.

6.5.3 Tranquillizers

(a) Benzodiazepines

The benzodiazepines are an important class of pharmacologically active

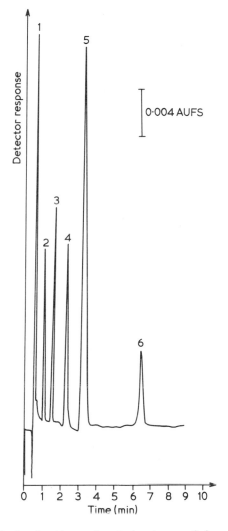

Fig. 6.11 The analysis of anticonvulsants in plasma. Column packing, LiChro-sorb SI 100 (10 μm); column dimensions, 250 x 2.1 mm i.d.; eluent, dichloro-methane/tetrahydrofuran (95:5), pressure, 967 psi (6.6MPa); flow rate, 1 ml min⁻¹; column temperature; ambient; detection, UV at 250 nm. Solutes: 1, solvent; 2, phenobarbitone; 3, phenytoin; 4, nitrazepam; 5, carbamazepine; 6, carbamazepine-10,11-epoxide. Reproduced with permission from reference 223.

compounds. They are among the most frequently prescribed drugs today and it is essential to have rapid, quantitative methods for their analysis. Several chromatographic procedures for the separation of benzodiazepines have been reported. These include partition [248, 318, 319], adsorption [298, 320, 321] and ion exchange chromatography [322]. Efficient separations are achieved in all these systems. Slight tailing of the peaks has been observed on silica using a mobile phase of 1 per cent methanol in chloroform [320] but good symmetrical peak shapes may be obtained by deactivation of the silica surface with water-saturated dichloromethane/isopropanol (96:4) [298].

Analysis of pharmaceutical products.   Diazepam, nitrazepam and oxazepan can be separated on a column of Durapak OPN/Porasil C with a mobile phase of isopropyl ether/tetrahydrofuran (90:10) [248]. Using UV detection at 254 nm, detection limits of 5 ng, 15 ng and 40 ng respectively have been reported. Rodgers has analyzed chlordiazepoxide (Librium) in capsules by dissolution of the material in 4 per cent diethylamine in chloroform containing 4-hydroxy-acetanilide as internal standard, prior to chromatography [320]. Separation was effected on SIL-X-II pellicular silica with a mobile phase of chloroform/methanol (98.5:1.5). The chlordiazepoxide was quantified by determination of peak height ratios at 254 nm. Good precision was observed.

   The analysis of the benzodiazepine ketazolam is complicated by the fact that the drug slowly degrades to diazepam in non-aqueous systems. GLC is not a suitable technique for distinguishing the two compounds since ketazolam is immediately pyrolized to diazepam. An HPLC separation has been achieved on a 1 m column of Corasil II with a mobile phase of isopropyl ether/tetrahydrofuran (85:15) [248]. There was only about 0.5 per cent loss of ketazolam during the 20 min required for the analysis, making the method suitable for most determinations. Using detection at 254 nm, around 5 ng of diazepam and 30 ng of ketazolam can be detected. This procedure has been used to study the kinetics of the decomposition of ketazolam in non-aqueous solution.

Analysis in body fluids   An ethyl acetate extract of urine from a dog which had been treated with the benzodiazepine I (Fig. 6.12) was analyzed on a 1 m x 2 mm i.d. column of Durapak OPN [319]. As well as unchanged drug, the sample contained the metabolite II. These compounds were partially separated by elution with hexane followed by hexane/isopropanol (70:30). Diazepam and N-desmethyldiazepam can be determined in blood by chromatography on Partisil 10 with detection at 232 nm [323]. The samples were eluted with heptane/isopropanol/methanol (40:10:1). 1.0 ml of blood was sufficient for a quantitative determination of the benzodiazepines in concentrations above 100 ng ml^{-1}. For the determination of lower levels 2.0 ml of blood and a more elaborate extraction procedure were necessary.

   Triflubazam and its primary metabolites have been determined in blood and urine by adsorption chromatography on a pellicular silica with isooctane/dioxan (85:15) as eluent [321]. Quantitative analysis of the drug and its metabolites was obtained using an internal standard, 11 α-OH, 17 α-methyl-testosterone and UV detection at 254 nm. Recoveries for all the compounds averaged about 95 per cent and the method was sensitive to 50 ng ml^{-1}. The specificity of the method has been established by collecting the eluted samples and characterizing them by mass spectrometry.

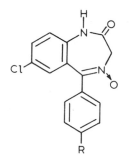

	R
I	H
II	OH

Fig. 6.12   Structure of a representative benzodiazepine (I) and its metabolite- (II).

A reversed phase procedure which is applicable to the determination of several benzodiazepines and their hydrolysis products in blood and urine has also been described [324].

(b)   Phenothiazines

Knox and Jurand have studied the separation of several phenothiazines as well as other common tricyclic psychosedative drugs by ion pair partition and adsorption chromatography [295]. They concluded that the latter method was more selective for individual members of a single class and this is the approach which has been used by most investigators [298, 325, 326]. The separation of oome oulphur containing tricyclic drugs has been investigated on Carbowax 400 with chloroform/isopropanol (98:2), but incomplete resolution was obtained [318].

Phenothiazines with piperazine side chains are highly polar and therefore difficult to analyze by GLC. Rodgers has studied the chromatography of several phenothiazines including those with piperazine side chains, by adsorption and ion exchange chromatography [320]. Not all of the phenothiazines studied could be completely resolved with either system.

Phenothiazine formulations have been analyzed by chromatography on Vydac reverse phase material with a methanol/water (50:50) mobile phase [327]. Using o-phenylphenol as internal standard, and UV detection at 254 nm, a coefficient of variation of 0.35 per cent was obtained for the determination.

The liquid chromatographic behaviour of oxidized homologues of the thioridazine and northioridazine series has been investigated [325, 328], on microparticulate silica gel (LiChrosorb SI 60). Variation of the ammonia concentration between 0.5 and 1.5 per cent in diisopropyl ether/isopropanol (85:15) permitted the adjustment of the solvent system to fit the different basicities of the various groups of interest [325]. With UV detection, the peaks for thioridazine ring oxide (which does not fluoresce) and mesoridazine (which is strongly fluorescent) were detected as poorly resolved peaks, but with fluorescence detection, only the mesoridazine peak was detected [328]. For the quantitative determination of both unresolved peaks, the ratio of the areas of the

UV and fluorescence peaks of mesoridazine itself could be computed. The peak area of the mesoridazine UV peak was obtained by difference, and thus the peak area of the thioridazine ring oxide was also found. The coefficient of variation ($n = 8$) for the determination of ring oxide was 3.5 per cent for injections of <0.1 μg. The coefficient of variation for the mesoridazine determination via the fluorescence signal was <2 per cent with a detection limit of 6 μg.

Hence it can be useful to have the use of combined UV and fluorescence detectors for the analysis of poorly resolved pharmaceutical compounds only one of which is fluorescent, or for the determination of traces of a highly fluorescent component in the presence of a large excess of a non-fluorescent substance. In both cases, it is difficult to obtain satisfactory results with only one detection system. However, it must be remembered that the presence of a large UV absorbing endogenous peak may cause some quenching of the fluorescence, resulting in low estimates of the fluorescent material present.

Analysis in body fluids. In an HPLC analysis of thioridazine and five metabolites in blood shown in Fig. 6.13, Muusze and Huber developed an on-line fluorimetric microreaction detector for the selective detection of fluorophores formed by oxidation of the phenothiazines [50]. The oxidation was carried out using excess permanganate/acetic acid in a 1.4 m length of 0.4 mm i.d. tubing (reaction time 5 s) and the excess permanganate was destroyed with hydrogen peroxide in 100 mm of the same tubing. The total volume of both reactors including mixing heads was 200 μl, the volume being kept as low as possible to avoid loss of resolution. Plasma extracts were chromatographed on silica using mobile phases comprising isooctane and 2-aminopropane with additions of acetonitrile and ethanol. The 2-aminopropane was thought to compete with the phenothiazines for the acidic silanol sites and thus, by increasing the concentration of the amine in the mobile phase, the capacity ratios of the drugs were reduced. Addition of ethanol or acetonitrile had a similar effect. In this case fluorimetric and UV detection gave similar detection limits on test mixtures of the compounds although fluorimetry is expected to be more useful in general for the analysis of biological samples due to its higher selectivity. A large UV absorbing endogenous peak which masked the thioridazine peak in the chromatography of plasma extracts was not observed using the fluorimetric reaction detector, and hence quantitation of the thioridazine could be accomplished. However, due to the larger volume of mixing in this system, the resolution between thioridazine and northioridazine was poorer.

A sample of chlorpromazine in blood serum (at an estimated concentration of 1.4 μg in a 5 μl injection) was analyzed by direct injection onto a strong cation exchange column and elution with 0.01M ammonium phosphate buffer at pH 9 [320]. It was estimated that as little as 5 ng could be detected under optimum conditions, using a UV detector at 254 nm.

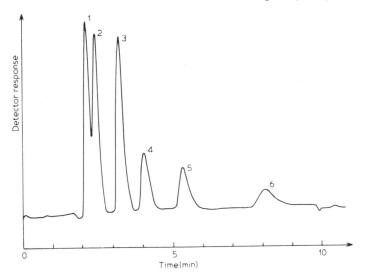

Fig. 6.13 The analysis of thioridazine and its metabolites in blood. Column packing, silica (9 μm); column dimensions, 250 x 2.8 mm i.d.; eluent, isooctane/ acetonitrile/2-aminopropane (95.87:2.69:0.96); pressure, 2 000 psi (13.6 MPa); flow rate, 1.14 ml min^{-1}; column temperature, ambient; Fluorimetric detection, excitation, 365 nm, emission, 440 nm. Solutes 1, thioridazine; 2, northio-ridazine; 3, thioridazine-2-sulphone; 4, thioridazine-2-sulphoxide; 5, thiorid-azine-5-sulphoxide; 6, northioridazine-2-sulphoxide. Reproduced with permission from reference 50.

6.6 Analgesic and anti-inflammatory drugs

6.6.1 Analgesics
Several early papers dealt with the HPLC analysis of a range of common analgesics. The quantitative analysis of mixtures of aspirin (acetylsalicylic acid), caffeine, paracetamol (*p*-hydroxyacetanilide or *p*-acetaminophenol), phenacetin (*p*-ethoxyacetanilide) and salicylamide was carried out on a 3 m column of LFS pellicular anion exchange resin with a mobile phase of 1.0*M* Tris at pH 9.0 [329]. The active components in 12 commercial analgesic tablets were analyzed. A similar range of compounds has been analyzed on cation or anion exchangers based on polystyrene [330], or microparticulate silica [331]. One of the earliest papers on the analysis of analgesics by HPLC was that of Henry and Schmit where various combinations from aspirin, caffeine, paracetamol, phenacetin and salicylamide in six types of tablets were analyzed on a 1 m x 2.1 mm column of Zipax SAX with an aqueous mobile phase buffered at pH 9.2 [237]. Sample preparation simply involved dissolution of the tablet in methanol and injecting an aliquot onto the column. In a tablet containing only paracetamol, the minimum quantifiable amount was 3 ng (using salicylamide as internal standard). Free salicylic acid could also be chromatographed on this column using a buffer

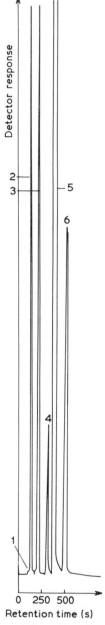

Fig. 6.14 The analysis of a multicomponent analgesic product. Column packing, μBondapak C_{18} (10 μm); column dimensions, 300 x 4 mm i.d; eluent, 0.01 per cent aqueous ammonium carbonate/acetonitrile (60:40); flow rate, 1.0 ml min^{-1}; column temperature, ambient; detection, UV at 254 nm; sensitivity, 0.16 AUFS. Solutes: 1, salicylic acid; 2, aspirin; 3, caffeine; 4, butalbarbitone; 5, phenacetin; 6, *p*-chloroacetanilide. Reproduced with permission from reference 334.

solution of higher ionic strength, and its presence as an impurity was demonstrated in a freshly dissolved analgesic tablet, the concentration increasing as the active ingredients slowly decomposed in solution [237]. Free salicylic acid can also be determined on Corasil II with a mobile phase of dichloromethane/acetic acid (99:1) [332]. It is important to be able to detect salicylic acid in aspirin preparations since the level is controlled as described in the British Pharmacopaeia. Paracetamol, salicylamide and aspirin have also been detected polarographically after chromatography on a column of AS-Pellionex-SAX with a mobile phase of 0.005M sodium nitrate at pH 9 [333]. Trace levels of p-aminophenol could be detected in paracetamol preparations with an electrochemical detector, following cation exchange chromatography [249]. The resolution and speed obtainable in modern HPLC is well illustrated in Fig. 6.14 by the separation of salicylic acid, aspirin, phenacetin, caffeine, butalbarbitone and p-chloroacetanilide (internal standard) in 10 min [334]. No loss in column performance was experienced even after 1000 injections. An automated system has been described for the separation and quantitation of aspirin, phenacetin and caffeine, allowing one analysis every 7 min [332]. Chromatography was performed on a controlled pore glass support with a chloroform/acetic acid (92:8) mobile phase and detection was at 280 nm. Methaqualone and paracetamol have been found difficult to resolve on silica using chloroform/methanol mixtures as eluent [335]. Adjusting the polarity of the mobile phase did not improve the separation; however, by changing to chloroform/isopropanol mixtures and making use of secondary solvent effects (p. 43) complete separation of the two peaks was obtained in 3 min [335].

Significant on-column decomposition of propoxyphene was observed during analysis by GLC [336], therefore an assay using HPLC was developed. Only with microparticulate silica columns (10 μm) could propoxyphene be completely resolved from both thermal degradation products and decomposition products formed in the presence of paracetamol. The mobile phase which gave the best combination of resolution and minimum analysis time was a basic isopropanol/hexane mixture. Coefficients of variation of 1.56 per cent and 1.96 per cent were obtained for the determination of seven tablet and capsule sample composites respectively [336].

6.6.2 Cough-cold preparations

The analysis of multicomponent pharmaceutical dosage forms is difficult. GLC has been used but the free bases usually present in these mixtures cause problems unless they are derivatized. Spectrophotometric methods are usually not specific enough without an initial chromatographic step.

Honigberg *et al.* have investigated the qualitative and quantitative analysis of 21 drugs which are common components of cough-cold mixtures [337]. Most compounds were well resolved on a column of Phenyl/Corasil (37–50 μm) using a mobile phase of acetonitrile/1 per cent ammonium acetate at pH 7.4. This pH was found to be the optimum for chromatographic efficiency. Several active ingredients of antitussive syrups have also been separated by adsorption chromatography on Spherosil [335]. However, neither of these methods has actually been applied to the analysis of pharmaceutical samples.

Paracetamol, dextromethorphan hydrobromide and chlorpheniramine maleate have been quantitatively analyzed in antitussive syrups by chromatography on Permaphase ODS with a methanol/buffer solution [338]. Problems were experienced because of the large excess of paracetamol present in the syrups but these were overcome by determining the minor components separately after extraction into chloroform. Two or three component mixtures from decongestant-antihistamine combinations have been separated on Zipax SCX using a phosphate buffer/dioxan mobile phase [339]. The ingredients which were analyzed included phenylpropanolamine hydrochloride, chlorpheniramine maleate, pseudoephedrine hydrochloride, pheniramine maleate and mepyramine maleate.

6.6.3   Anti-inflammatories

The adrenocortical steroids hydrocortisone, cortisone, hydrocortisone acetate and cortisone acetate may be separated by liquid–liquid chromatography on a cyanoethylsilicone column using water/methanol (99:1) as eluent [236]. This method has been applied to the assay of these steroids in various dosage forms, primarily creams and ointments. The sample preparation merely involved dissolution of the steroid in alcohol plus the addition of an internal standard prior to injection. Excipients were usually unretained, but if large amounts were chromatographed they could interfere with the hydrocortisone determination. The results of a study of the controlled decomposition of hydrocortisone in basic solution indicated a good separation of the drug from its degradation products [236]. The simultaneous analysis of hydrocortisone and hydrocortisone phosphate in an injectable dosage form has been achieved by reversed phase ion pair partition chromatography [250]. The mobile phase consisted of methanol/water containing tetrapentylammonium hydroxide at pH 7.5. UV detection at 254 nm was used. Since hydrocortisone phosphate was present in a large excess, two internal standards (p-propenylanethole and benzophenone) were incorporated into the mixture. The detector was attenuated after elution of the benzophenone and before p-propenylanethole and hydrocortisone phosphate eluted. Preparations containing 0.05 per cent of hydrocortisone could be analysed by this method. Lignocaine, hydrocortisone acetate and phenylbutazone could be separated on a 150 x 4.8 mm column of Spherosil using a gradient of ethanol in cyclohexane [335]. These components occur together in anti-inflammatory preparations but the method was not applied to the analysis of pharmaceuticals.

Several synthetic corticosteroids have been assayed in pharmaceutical preparations by HPLC [107, 235, 340]. Fluocinolone acetonide has been analyzed in creams and fluocinolone acetonide acetate in ointments by chromatography on Permaphase ODS at 40°C using water/methanol (92.5:7.5) and (70:30) respectively [235]. Toluene was used as the internal standard in both cases. Precolumn preparation simply required partition of the excipients into isooctane and the active agent into aqueous alcohol. Triamcinolone acetonide, used topically in the treatment of skin diseases, is often blended with rice starch before individual formulations are prepared. A quality control procedure for the blended powders consisted of extraction of the steroid into dichloromethane followed by chromatography on a column of Zipax using water-saturated dichloromethane as eluent and UV detection [340].

Dexamethasone disodium phosphate can be determined in tablets by ion exchange chromatography on Zipax WAX at 40° C using a mobile phase of water/ethanol (90:10) [107].

6.6.4 Analysis in body fluids

Phenylbutazone has been analyzed in plasma by HPLC on a 1 m x 1.8 mm i.d. column of SIL-X-adsorbent, coiled into a radius of 150 mm [342]. The mobile phase was hexane/tetrahydrofuran/glacial acetic acid (90:10:0.002) and a column temperature of 35° C was used. The phenylbutazone was extracted with hexane from acidified plasma to which an internal standard (benzaldehyde-2,4-DNP) had been added. No extraneous peaks were observed and the analysis was accomplished in 7 min. Two known metabolites of phenylbutazone, hydroxy-phenylbutazone and oxyphenbutazone were not observed in any of the biological samples analyzed, although it is possible that the more polar metabolites were not extracted from the plasma by hexane. Phenylbutazone concentrations of 1.75 μg ml^{-1}, corresponding to 350 ng per injection, were determined at 254 nm. Recoveries from spiked plasma samples of 99.9 ± 1.7 per cent and 99.3 ± 3.1 per cent were obtained using electronic integration and peak height measurements respectively [342]. The potential of the method for metabolic studies was demonstrated by obtaining the plasma phenylbutazone profile of a male who had been administered two 100 mg tablets of the drug. Subsequent modification of the mobile phase to hexane/tetrahydrofuran/acetic acid (77:23:0.002) allowed simultaneous analysis of phenylbutazone and oxyphenbutazone in human plasma [343]. Using a UV monitor set at 254 nm the minimum detectable amount of phenylbutazone was 15 ng injected on-column while 20 ng of oxyphenbutazone could be detected.

Other applications of HPLC to the study of metabolic processes have been illustrated by work on the analysis of paracetamol and its metabolites in body fluids [344–347]. HPLC analyses of urine and serum samples from two clinically normal men who had ingested about 2 g of paracetamol gave separations of the free paracetamol and seven metabolites, viz., 2-methoxy paracetamol and its glucuronide and sulphate conjugates; the sulphate conjugate of 2-hydroxy paracetamol; the glucuronide and sulphate conjugates of para-cetamol; S-(5-acetamido-2-hydroxyphenyl)cysteine and S-(5-acetamido-glucuronosidophenyl) cysteine [344]. The latter metabolite, which had not been previously isolated, was characterized by mass spectrometry. Both UV and cerium fluorescence detectors were used to follow the excretion of the drug and its metabolites. However, this method has several disadvantages since the analysis was performed at 70° on an anion exchange resin using gradient elution and required 21 h. Faster methods have now been developed for the simultaneous estimation of paracetamol and its sulphate, glucuronide, cysteine and mercapturic acid conjugates in urine [345, 347] and plasma [345, 346]. Samples can be analyzed by normal phase [346], reversed phase [345] or soap chromatographic [347] methods. Using UV detection at 257 nm a concentration of 1 μg ml^{-1} paracetamol in 1 ml urine samples could be detected [346]. Riggin *et al.* have determined paracetamol in plasma and urine by chromatography on a column of Pellidon polyamide followed by electrochemical detection [249]. The limit of detection in serum was 150 ng but only 5 μg could be detected in urine samples because of interference from uric acid.

Phenacetin has been determined in the blood, urine and tissues of treated dogs by chromatography on a column of Porasil A using isooctane/dioxan (70:30) or chloroform/dioxan (90:10) as mobile phase [348]. The limit of detection for phenacetin in plasma or urine was 0.1 μg ml^{-1} and 0.5 μg ml^{-1} in tissue.

Analysis of *p*-aminobenzoic acid and its metabolites in human urine and serum has been achieved on a column of AS-Pellionex SAX using a formate buffer at pH 3.5 as mobile phase [238]. The quantitative determination of *p*-aminobenzoic acid, *p*-aminohippuric acid, *N*-acetyl-*p*-aminobenzoic acid and *N*-acetyl-*p*-aminohippuric acid was possible. Urine and serum samples could be analyzed directly without extraction or pretreatment using 2 μl injections and all four compounds eluted within 20 min. The detection limits were 5 ng (signal: noise = 3:1) at 254 nm. The method was used for a pharmacokinetic study and was shown to be suitable for routine clinical analysis.

In the conventional determination of indomethacin in body fluids by spectrofluorimetry, aspirin and frusemide interfere and must be removed by extraction or chromatography. The GLC assay involves derivatization of the drug, so a simplified method based on HPLC was developed [349]. Indomethacin and an added internal standard (flufenamic acid) were extracted from plasma and chromatographed on a column of μBondapak C$_{18}$/Corasil (37–50 μm) using 0.1M acetic acid/acetonitrile (60:40) as the mobile phase. The method was reproducible, quantitative and sensitive down to plasma concentrations of 0.1 μg ml^{-1}. Aspirin and frusemide eluted in the void volume and did not interfere.

The blood of patients treated with the anti-inflammatory drug, diftalone (I) (Fig. 6.15), contains unchanged I and two metabolites II and III. These have been determined on a column of Permaphase ODS at 54° C using dexamethasone alcohol as internal standard [350]. The analyses were either performed isocratically or under gradient conditions using aqueous acetonitrile mobile phases. The detection limit for all three compounds was 0.5 μg ml^{-1}.

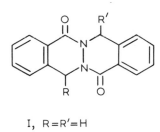

I, R=R'=H

II, R=OH; R'=H

III, R=R'=OH

Fig. 6.15   Structure of diftalone (I) and its major metabolites (II & III).

6.7   Diuretics

6.7.1   *Pharmaceutical preparations*

Honigberg *et al.* have studied the separation of seven common drugs often found in combination in diuretic-anti-hypertensive mixtures [351, 352]. Chromatography on both C_{18}/Corasil and Phenyl/Corasil was investigated with acetonitrile or methanol/water mobile phases containing ammonium salts [351]. The determination of a reserpine-chlorothiazide mixture required two stages since there was a large difference in the concentration of the two ingredients [352]. The reserpine was first extracted from the mixture with chloroform. The residue was resuspended and diluted with methanol for the chlorothiazide determination. Both samples could be chromatographed on C_{18}/Corasil using aqueous 0.5 per cent ammonium chloride at pH 5.6/methanol (90:10) with UV detection.

Powdered tablets of polythiazide have been extracted with methanol, containing quinoline as internal standard, and assayed by chromatography on a 600 x 3 mm i.d. column of Phenyl/Corasil using water/methanol (65:35) as the mobile phase [353]. Sharp, symmetrical peaks were obtained and the sensitivity at 254 nm was good enough for single tablet assays. It was also shown that the method was adaptable to the determination of many other thiazide diuretics and eleven other thiazides were examined. However, on repetition of this work, Cohen and Munnelly concluded that quinoline was not a suitable internal standard and that the presence of other UV absorbing material caused inaccuracies [354].

Canrenone has been analyzed in pharmaceutical preparations without interference from common excipients or degradation products [355]. The extracts were chromatographed on a 1 m x 2.1 mm i.d. column of SIL-X with a mobile phase of hexane/isopropanol (96:4) using *o*-nitroaniline as internal standard.

6.7.2. Determinations in body fluids

Frusemide, a widely used diuretic, and its hydrolysis product, 4-chloro-5-sulphamoylanthranilic acid, have been separated by high performance ion exchange chromatography and quantitated fluorimetrically [356]. Plasma and urine extracts were chromatographed on a column of Permaphase AAX using $0.008M$ phosphate buffer at pH 8.0 as mobile phase. Dilute hydrochloric acid was added to the eluate via a jet-flow mixer and the fluorescence of the acidified material determined. Concentrations of $2~\mu g~ml^{-1}$ of either compound were assayed in plasma with a coefficient of variation of 8 per cent and $0.1~\mu g~ml^{-1}$ could be assayed in urine with similar precision. Frusemide has also been determined in the plasma and urine of normal and uraemic patients by chromatography on a pellicular cation exchange resin, with an ammonium phosphate buffer at pH 2.5 as eluent [293]. Fluorimetric detection was used for the determination. $5~\mu l$ amounts of serum and urine were injected directly onto a precolumn which protected the main column from contamination.

2-Aminoethylhydrogen sulphate has been determined in biological fluids by cation exchange chromatography using an eluent of $0.05M$ citrate buffer at pH 2.6 [357]. Detection was accomplished with an automated fluorescamine detection system. The drug was assayed at concentrations of $0.3~\mu g~ml^{-1}$ using 2 ml aliquots of plasma or urine. Linear responses were observed up to $133~\mu g~ml^{-1}$ and the method was found to be accurate and reproducible.

A quantitative method has been developed for the analysis of hydrochlorothiazide in serum at the therapeutic level [358]. The method is based on exclusion chromatography of the sera on Sephadex G-15, extraction of the protein-free fraction of the effluent with ethyl acetate and injection of a methanol solution of the extract onto a reversed phase Spherisorb ODS column. Water/methanol (85:15) was used as the mobile phase and detection was at 280 nm. The detection limit was 50 ng.

The carbonic anhydrase inhibitor, acetazolamide can be determined in 1 ml plasma samples at the 25 ng level [359]. The drug and its propionyl analogue as internal standard were extracted from plasma with ethyl acetate, and the lipids removed with dichloromethane. The samples were then chromatographed on a column of ODS-silica with a mobile phase of $0.05M$ sodium acetate buffer at pH 4.5/methanol (98:2) and UV detection at 254 nm was used. The method was developed for the purpose of measuring the comparative bioavailability of acetazolamide in different formulations.

6.8 Drugs of abuse

Sensitive and rapid analyses of drugs of abuse are needed by law enforcement agencies and in toxicology and criminology laboratories. These analyses are important in determining the composition of seized substances and the quantity of drug in a specific sample. Since many of the samples are mixtures or impure, identification or quantitation by spectroscopic methods usually requires preliminary clean-up. The high polarity and low volatility of many of the drugs restrict the use of GLC, and TLC is of limited use for quantitative analyses. HPLC is inherently suitable for dealing with these samples.

6.8.1 Phenethylamines

A range of phenethylamines of forensic interest have been separated both on an anion exchange resin as their hydrochlorides and by adsorption chromatography on Corasil II as their free bases [360]. However, the anion exchange separations required fairly lengthy analysis times, even with elevated operating temperatures. Methamphetamine, methoxyphenamine and ephedrine were separated as their free bases at ambient temperature on Corasil II with a chloroform/methanol (80:20) mobile phase.

Improved analytical methods have been developed with the advent of small particle porous packings. Retention data have been reported for twenty seven amphetamine-type stimulants on a column of microparticulate silica using a mobile phase of methanol/$2N$ ammonia/$1N$ ammonium nitrate (27:2:1) [361]. The separation of benzphetamine, phendimetrazine, phenmetrazine, dex-amphetamine, N-methylephedrine, ephedrine, methylamphetamine and meph-entermine was achieved in 15 min using this system.

Ketamine (Fig. 6.16, I) is a parenteral anaesthetic agent which has recently been abused as a hallucinogen. Few methods were available for the detection of the drug and an HPLC method was developed for the analysis of ketamine and two of its *in vivo* metabolites (II and III) in rat urine [362]. After extraction from the urine, the drug and its metabolites were converted to p-nitrobenzyl derivatives since the underivatized materials exhibit low UV absorbance. These derivatives were separated on a column of μBondapak C_{18} with a water/acetonitrile mobile phase. By-products and excess reagents did not interfere in the analysis.

6.8.2 Opium alkaloids

Separations of opium alkaloids by liquid–liquid partition chromatography [363], adsorption chromatography [364] or ion exchange chromatography [365, 366] have been described. These methods are discussed in more detail under Alkaloids (Section 6.9) and only the applications of forensic interest are considered here.

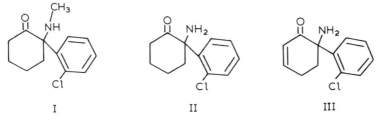

I	II	III

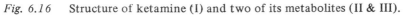

Fig. 6.16   Structure of ketamine (I) and two of its metabolites (II & III).

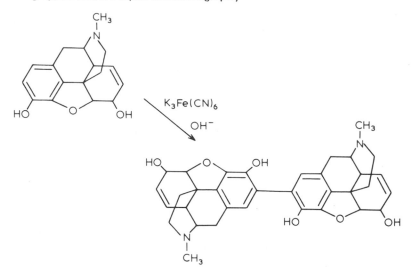

Fig. 6.17   The oxidative dimerization of morphine.

A highly specific and sensitive method developed for the analysis of morphine in urine, involved conversion of the alkaloid, after hydrolysis of the glucuronide conjugate present in urine, to the fluorescent dimer pseudomorphine under mild alkaline oxidative conditions as shown in Fig. 6.17 [367]. For greater quantitative accuracy, dihydromorphine was used as internal standard and the three dimeric products formed (pseudomorphine, bis-dihydromorphine and morphine-dihydromorphine) were separated by HPLC on a column of 7 μm Partisil using methanol/2N ammonium hydroxide/1N ammonium nitrate (30:20:10) as mobile phase. Fluorimetric detection was used with excitation and emission wavelengths of 320 nm and 436 nm respectively. Drugs such as normorphine and paracetamol which also undergo the oxidative coupling reaction could be distinguished from morphine by their relative retention times. A total of 53 common drugs were shown not to interfere with the method. For quantitation, a linear relationship was obtained between the amount of morphine in solution and the peak height ratio of the morphine-dihydro-morphine dimer to the bis-dihydromorphine dimer. Levels of 0.1–10 μg ml^{-1} morphine in urine could be quantitated and an absolute detection limit of 4 ng injected on column was obtained.

A rapid method of analysis of illicit heroin samples for forensic purposes has been developed by Twitchett [368]. No extraction or derivatization of the heroin was required and chromatography was performed on a column of Zipax SCX using gradient elution with a borate buffer. A separation of barbitone, caffeine, morphine, 6-O-acetylmorphine, strychnine, heroin (diamorphine), quinine and cocaine was obtained in 12 min using this system, as shown in Fig. 6.18. For quantitative analysis, diphenylamine was added as internal standard and peak height ratios were used for the determination. The detection wavelength of choice was 270 nm. Greater sensitivity may be achieved at 235 nm but baseline drift was experienced during the solvent programme. Use was made of retention time as an index for the tentative identification of constituents, the retention characteristics of fourteen possible components being recorded. Peak identification was confirmed by mass spectrometry after extraction of the eluent fractions with ether or chloroform. The analysis of 20 illicit heroin samples showed a total content of morphine drugs (heroin, 6-O-acetylmorphine, and morphine) of 40—65 per cent and, as well as caffeine, the major diluent, small amounts of strychnine (0—1.4 per cent) and quinine were often present.

Several opium alkaloids have been chromatographed on Partisil using a mobile phase of methanol/2N ammonium hydroxide/1N ammonium nitrate (27 : 2 : 1) [361]. Peak heights were used to quantify the components, a coefficient of variation of less than 2 per cent being obtained from replicate injections of a standard solution. The detection limit for morphine and heroin at 278 nm was 50 ng and a linear calibration graph was obtained for 0.05—10 μg morphine injected. The major components of an opium sample were found to be narcotine, thebaine, codeine and morphine, while analysis of a 'Chinese heroin' sample showed the presence of caffeine, heroin, monoacetylmorphine, morphine and strychnine. This procedure for the analysis of illicit heroin samples is much simpler than the previous one [368] since a gradient elution technique is not necessary to separate the components.

6.8.3 Cannabis samples

For law enforcement purposes, the identification of cannabis samples usually involves only chromatography or microscopy. It is sometimes necessary to compare cannabis samples to establish a common origin or to trace distribution chains, and TLC and GLC have been used for this purpose, although both these methods may lack discrimination. The potential of HPLC for comparative cannabis analysis has been investigated and it was found to be a superior technique to GLC and TLC, distinguishing 30 out of 34 samples of cannabis resin and herbs and failing only to distinguish different samples of the same geographical origin [369]. GLC distinguished 25 out of the 34 samples, and TLC 11 of the 34. The analyses were performed in 10—15 min on an ODS—bonded silica using a mobile phase of methanol/0.02N sulphuric acid (80 : 20). Solutions were prepared for injection by crushing the cannabis (resin or herbal) sample, extracting with mobile phase and injecting the supernatant solution onto the column. The acidic mobile phase was found to improve the peak shapes of longer retained components. The HPLC results were not affected by ageing the extracts for several days. No attempt was made to identify the components in the extract. In an earlier analysis of a hashish extract, preliminary size separation by exclusion chromatography was used to obtain a fraction of molecular weight corresponding to that of tetrahydrocannabinol and this fraction was rechromatographed on Permaphase ODS using a water/methanol gradient at a column

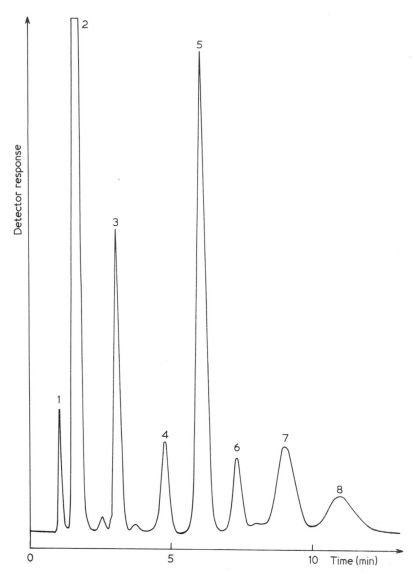

Fig. 6.18 Analysis of the constituents of an illicit diamorphine sample. Column packing, Zipax SCX; column dimensions, 1.2 m x 2.1 mm i.d.; eluent, linear gradient (in 6 min) from 0.2M boric acid, adjusted to pH 9.3 with 40 per cent hydroxide to 0.2M boric acid/acetonitrile/n-propanol (86:12:2) adjusted to pH 9.8 with 40 per cent sodium hydroxide; flow rate, 2 ml min^{-1}; column temperature, ambient; detection, UV at 270 nm; sensitivity, 0.2 AUFS. Solutes: 1, barbitone; 2, caffeine; 3, morphine; 4, monoacetylmorphine; 5, strychnine; 6, diamorphine; 7, quinine; 8, cocaine; Reproduced with permission from reference 368.

temperature of 50°C, but no further classification of the peaks was attempted [370]. More recently Smith has located cannabinol (CBN), cannabidiol (CBD), Δ^8- and Δ^9-tetrahydrocannabinol (Δ^8- and Δ^9-THC) in the liquid chromatogram of a cannabis extract by comparison with standards [371]. The identity of these components was confirmed by examination of the fractions by GLC and GLC-MS. The HPLC conditions were the same as those used for the comparative analyses mentioned above [369]. An acidic extract of cannabis was found to contain four major components on examination by HPLC [371]. After decarboxylation, the major products were identified as CBN, CBD and Δ^9-THC indicating that the acidic fraction contained cannabinolic acid (CBNA), cannabidiolic acid (CBDA) and Δ^9-tetrahydrocannabinolic acid (Δ^9-THCA) respectively. These samples could not be analyzed by GLC without prior derivatization [369].

The separation of the dansyl derivatives of Δ^9-THC, CBD and 11-hydroxy-Δ^8-THC has been achieved in less than 20 min on a column of MicroPak SI-10 using gradient elution from hexane to 2 per cent methanol in dichloromethane [372]. Both the mono- and di-dansylate peaks were obtained for CBD. Better sensitivity and baseline stability was obtained with fluorescence detection than with UV absorbance at 254 nm in the chromatogram of 30 ng of dansyl-11-hydroxy-Δ^8-THC. The signal-to-noise ratio of the fluorescence peak was 600 times that of the absorbance peak and the fluorescence detection yielded an extrapolated minimum detectable quantity of 5 pg (signal : noise = 2) compared with 3 ng using UV detection. This sensitivity would be suitable for the analysis of physiological fluids containing THC related compounds.

6.8.4 Lysergic acid diethylamide (LSD)

An analysis of LSD in illicit preparations for forensic purposes has been developed by Jane and Wheals [373]. A range of column packing materials was examined including C_{18}/Corasil, Corasil II, OPN Porasil C, Carbowax 400, Porasil C and Zipax SCX. Problems of reproducibility with the adsorbents led to a preference for the reversed phase material, C_{18}/Corasil, and a mobile phase of methanol/0.1 per cent ammonium carbonate (60 : 40) was used. The ammonium carbonate in the mobile phase gave improved resolution and shortened the retention times. Although the sensitivities of the UV and fluorescence detectors were similar for the detection of LSD (with lower limits of detection of LSD around 2–10 ng), fluorimetric detection was found to be more selective for the ergot alkaloids, and interferences from excipients in LSD tablets, apparent when UV detection was used, were eliminated. Using the fluorimetric detector a linear relationship between peak height and the amount of LSD injected was obtained over the range 0–75 ng. A satisfactory analysis of LSD was obtained in 50 different forms of LSD tablets. For HPLC analysis, half of a tablet was crushed in methanol (0.5 ml) and 5 μl of the supernatant solution was injected onto the column. The LSD peak was collected, diluted with water and its fluorescent characteristics determined. Although the excitation and emission wavelengths of

320 nm and 418 nm respectively are similar to those of other ergot alkaloids, the selectivity of the chromatographic separation was thought to be sufficient to characterize the material as LSD. The retention times of a range of ergot alkaloids measured relative to LSD were as follows: iso-LSD (1.39), ergometrine maleate (0.74), ergocryptinine base (2.28) and ergotamine tartrate (1.70). GLC was found to be less satisfactory for the analysis since high temperatures were required to chromatograph the silylated LSD, and in many cases interfering peaks were observed in the chromatogram.

Several other methods involving HPLC have been reported for the analysis of LSD (361, 374–376). However, in general these methods are less sensitive. A combination of HPLC and TLC has been found to be satisfactory for the identification of LSD in biological fluids, especially urine [377]. Urine extracts were chromatographed on a column of Partisil (6 μm) using an eluent of methanol/0.2 per cent ammonium nitrate (55:45). Fluorescence detection was used with an emission wavelength of 430 nm and an excitation wavelength of 325 nm. Fractions having the same retention times as LSD and iso-LSD were collected and individually analyzed by TLC along with standards. The samples were visualized first as blue fluorescent spots under irradiation with UV light at 360 nm and then by spraying with p-dimethylaminobenzaldehyde. The identity of the compounds could be further confirmed by mass spectrometry. The amounts of LSD isolated from the specimens examined ranged from 0.3–19.5 ng ml^{-1} and the efficiency of the extraction procedure was approximately 70 per cent.

6.8.5 Miscellaneous

The separation of sixteen common street drugs including phencyclidine (PCP), methadone, cocaine, tetrahydrocannabinol (THC), methylamphetamine, 2,5-dimethoxy-4-methylamphetamine (STP), methylenedioxyamphetamine (MDA), heroin, N,N-dimethyltryptamine (DMT), lysergic acid diethylamide (LSD), diazepam (Valium), mescaline, secobarbitone, amylobarbitone, phenobarbitone and diphenylhydantoin (Dilantin) has been achieved in 40 min on a 1 m x 2.3 mm column of Corasil II using a gradient system of ethanol/dioxan/cyclohexylamine in hexane [378]. The use of microparticulate porous silica should greatly improve the efficiency of this system. A wide range of drugs of abuse could be separated by isocratic elution on a 250 x 4.6 mm column of Partisil 'fines', obtained by sedimentation of 6 μm silica in methanol [361]. The solvent system used consisted of methanol/2N ammonia/1N ammonium nitrate (27:2:1), and at a flow rate of 1 ml min^{-1} (1500 psi; 10 MPa), an efficiency of 18 000 plates was obtained for methylamphetamine ($k' \sim 2$).

6.9 Alkaloids

Most alkaloids are pharmacologically active and many are of therapeutic value. Examples of the analysis of pharmaceutical constituents, including alkaloids, by HPLC have been discussed. In the present section the application of HPLC to the separation and analysis of alkaloids according to their class is considered.

6.9.1 Opium alkaloids

The development of rapid, sensitive, quantitative methods for the analysis of narcotics is extremely important to forensic scientists and toxicologists. HPLC has been applied to the analysis of illicit heroin samples [361, 368] and to the determination of morphine in opium samples [361] and urine [367]. These applications have been fully discussed in Section 6.8.

One of the earliest reported separations of opium alkaloids by HPLC was on columns of Zipax or Corasil coated with Poly G-300 using a mobile phase of heptane/ethanol (10:1) [363]. The separation on Zipax was incomplete but papaverine, thebaine, codeine and morphine were completely resolved in less than 30 min on Corasil using a dynamic coating technique with the eluent 40 per cent saturated with Poly G-300.

Morphine has also been separated from codeine, papaverine, thebaine, cryptopine and narcotine on a column of Zipax SCX with a mobile phase consisting of 0.2M sodium borate buffer at pH 9.6, 0.2M potassium nitrate/acetonitrile/n-propanol (95:4:1) [365]. Morphine itself had a fairly low k' value on the cation exchange column with an aqueous buffer mobile phase and the organic solvents were added to speed up the elution of the other alkaloids. Heroin can also be separated from its two hydrolysis products 6-O-acetylmorphine and morphine on the same column with a similar mobile phase. The synthetic analgesic, methadone could be separated from heroin and morphine on a column of Zipax SAX with a mobile phase buffered at pH 9.8 although under these conditions the two alkaloids were not completely resolved [365].

In optimizing the separation of morphine, heroin and 6-O-acetylmorphine on Zipax SCX and also that of methadone from the morphine group on Zipax SAX, Knox and Jurand have studied the effect of the pH and the ionic strength of the mobile phase, and the presence of organic modifiers in the mobile phase on the k' values of the alkaloids and on their plate heights [366]. As expected from ion exchange theory, log k' fell linearly with pH for morphine, 6-O-acetylmorphine and heroin on Zipax SCX, and k' was inversely proportional to the ionic strength. The addition of organic components to the mobile phase was found to be beneficial to the separation on Zipax SCX. Propanol (1–2 per cent) reduced k' and H more effectively than acetonitrile or methanol, but had to be used in small proportions (<3 per cent) to avoid degrading the polymeric film of the ion exchange resin. Higher concentrations of acetonitrile (15 per cent) were used without ill effects. The optimum separation of morphine, 6-O-acetylmorphine and heroin on Zipax SCX was obtained with a mobile phase of 0.04M sodium hydroxide buffered at pH 9.3 with boric acid/acetonitrile/n-propanol (86:12:2). At a pressure of 1600 psi (10.1 MPa) the separation was achieved in 30 min. Sample identity was confirmed by off-line mass spectrometry of collected fractions.

Verpoorte et al. have measured retention data for several alkaloids, including morphine, codeine, thebaine, heroin and papaverine, on LiChrosorb SI 60 (5 μm) using six different solvent systems comprising chloroform/methanol or diethyl ether/methanol combinations [364]. Some tailing of the alkaloid peaks was observed and poor efficiency, by present day standards, was obtained. However, the separations could probably now be improved with modern column packing techniques.

A method for the quantitative determination of thebaine in poppy plants using HPLC has been developed [379]. After extraction at pH 2, preliminary chromatography on Amberlite XAD-2 removed any non-alkaloid material. The thebaine was then separated from isothebaine and orientalidine on a column of Corasil II using a mobile phase of hexane/chloroform/methanol/ diethylamine (900:75:25:0.1). Quantitation was effected by peak height measurements and with 1 g poppy samples, 0.01 per cent of thebaine could be readily determined (detection limit of less than 25 ng), using a UV detector at 254 nm.

6.9.2 Tropane alkaloids

Tropane alkaloids have been examined by adsorption [364, 380, 381] reversed phase [382] and ion pair partition [383] chromatography.

Partial resolution of a mixture of scopolamine hydrobromide, eucatropine hydrochloride, tropine, apoatropine hydrochloride, homotropine sulphate, atropine sulphate and hyoscyamine hydrochloride was achieved on a 1 m x 4.5 mm column of SIL-X using a mobile phase of tetrahydrofuran/28 per cent ammonium hydroxide (100:1) [380]. However, complete separation of scopolamine, apoatropine and atropine have been obtained in 5 min on a 300 x 4.6 mm i.d. column of Partisil (5 μm) using a mobile phase of diethyl ether/methanol (90:10) plus 1 per cent diethylamine [381]. Honigberg et al. have analyzed scopolamine and atropine in antispasmodic mixtures by reversed phase chromatography on columns of C_{18}/Corasil or Phenyl/Corasil [382]. A mobile phase of methanol/water buffered with ammonium dihydrogen orthophosphate (1 per cent) or ammonium carbonate (0.5 per cent) was used. The concentration levels of scopolamine and hyoscyamine detected in this study were 100 times higher than the level commonly found in commercial dosage forms, but neither compound could be detected at therapeutic dose levels. The poor UV absorption of the tropane alkaloids is one of the major problems in their analysis by HPLC. Stutz and Sass obtained detection limits for the alkaloids which they examined of 50 μg with RI detection and 1 μg with UV detection [380]. Greatly improved detection limits have been obtained by chromatographing the alkaloids as ion pairs with picric acid [383]. Hyoscyamine, scopolamine and ergotamine were separated as picrate ion pairs on 100 mm columns of silica loaded with 0.06M picric acid in pH 6 buffer using a mobile phase of chloroform saturated with stationary phase as shown in Fig. 6.19. Close control of pH and temperature was found to be essential for reproducible separations. This approach provides a rapid, sensitive method of analysis of pharmaceutical products. The quantitative determination of scopolamine (6.8 per cent) in an excess of hyoscyamine (90.7 per cent) in an ampoule solution could be obtained in 8 min.

6.9.3 Indole alkaloids

(a) Miscellaneous indole alkaloids

Various chromatographic supports were examined for the small scale preparative separation of a mixture of indole alkaloids isolated from *T. holstii* [384]. The most suitable systems of those tested were Porasil C with a mobile phase of chloroform/methanol/ammonia (99:1:0.2), C_{18}/Porasil B using methanol or aqueous methanol containing ammonia as eluent, or alumina with a mobile phase of chloroform/hexane (90:10). When 100 mg of the plant extract was chromatographed on a 2.4 m x 12.5 mm column of Porasil C the first fraction still contained a mixture of compounds but the second fraction was found to be pure and crystallized on removal of the solvent to give pericyclivine.

The separation of the spiro isomers of the oxindole alkaloids from the genus *Mytragyna* has been achieved by reversed phase chromatography on C_{18}/Corasil at 60° C with methanol/water (80:20) as mobile phase [385]. It was also possible to distinguish between isomitraphylline, isorhynchophylline, uncarine F and isoteropidine.

(b) Ergot alkaloids

Several applications of HPLC to the analysis of LSD have been developed [361, 373–377]. These have been discussed in Section 6.8.4.

A procedure for the quantitation of ergotaminine (a pharmacologically inactive isomer of ergotamine) in plasma by HPLC with fluorescence detection has been published [386]. Chromatography was performed on a 250 x 2.2 mm i.d. column of silica with a mobile phase of isopropyl ether/acetonitrile/methanol (69.5:30:0.5). Using a fluorescence monitor detection limits of below 1 ng ml^{-1} were achieved for ergotaminine extracted from plasma.

Reversed phase chromatography on a 300 x 4 mm i.d. column of μBondapak C_{18} using a solvent gradient of $0.01M$ ammonium carbonate in water/acetonitrile has been used for testing the purity of ergotamine as an active substance and for checking its concentration in pharmaceutical preparations [387]. The system could be used for the identification and quantitative determination of seven known isomerization and hydrolysis breakdown products in a single run in less than 30 min, as shown in Fig. 6.20. Simultaneous UV detection at 280 and 320 nm also made it possible to measure further breakdown products formed by addition at the 9,10 double bond (lumi-compounds). The detection limit for all of the hydrolysis products was ~10 ng when 25 μg (50μl) of ergotamine tartrate was injected. There was no evidence of chromatographic disturbance by the excipients when this amount was injected.

Peak height measurements were used for quantitation. The precision of the determination was good with coefficients of variation for the determination of ergotamine and ergotaminine being 1.2 per cent and 1.0 per cent respectively ($n = 8$).

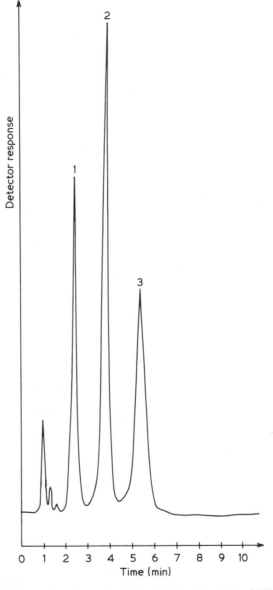

Fig. 6.19 Separation of tropane alkaloids. Column packing, LiChrosorb SI 100 (5 μm) loaded with 0.06M picric acid in buffer at pH 6; column dimensions, 100 × 3 mm i.d.; eluent, chloroform saturated with 0.06M picric acid in buffer at pH 6; flow rate, 0.6 ml min^{-1}; column temperature, ambient detection, UV at 254 nm. Solutes: 1, hyoscyamine (0.9 μg); 2, ergotamine (0.9 μg); 3, scopolamine (2.8 μg). Reproduced with permission from reference 383.

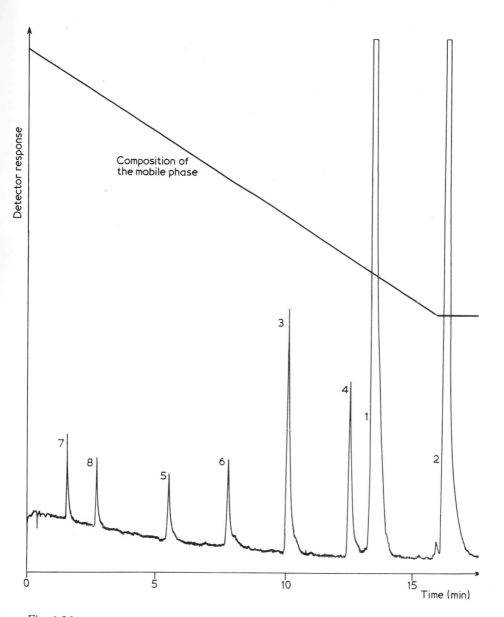

Fig. 6.20 Analysis of ergot alkaloids. Column packing, μBondapak C_{18} (10 μm); column dimensions, 300 x 4 mm i.d.; eluent, linear gradient from 0.01M aqueous ammonium carbonate/acetonitrile (37:3) to 0.01M aqueous ammonium carbonate/acetonitrile (50:50) in 15 min; pressure, 3000–3600 psi (20.4–24.5 MPa); flow rate, 8.0 ml min^{-1}; column temperature, ambient; detection, UV at 320 nm; sensitivity, 0.02 AUFS. Solutes: 1, ergotamine (9 μg); 2, ergotaminine (7 μg); 3, aci-ergotamine; 4, aci-ergotaminine; 5, lysergic acid amide; 6, isolysergic acid amide; 7, lysergic acid; 8, iso-lysergic acid (all 1–2 μg). Reproduced with permission from reference 387.

(c)   Strychnos alkaloids

While investigating the analysis of various alkaloid classes on silica gel using neutral solvent systems, Verpoorte and Svendsen found that alstonine and serpentine were not eluted within a reasonable period of time and strychnine and brucine were not separated [364]. The addition of 1 per cent diethylamine to the neutral solvent systems resulted in better separations by reducing the tailing and shortening the retention times [388]. A good separation of alstonine and serpentine was obtained in 10 min on a 300 x 2 mm i.d. column of containing 1 per cent diethylamine. Similarly strychnine and brucine were separated in 3 min using diethyl ether/methanol (90:10) containing 1 per cent diethylamine as eluent. Subsequently this approach was used in the analysis of twelve *Strychnos* alkaloids by HPLC [389]. The complete resolution of α- and β-colubrine could not be obtained by GLC or TLC but they were well separated LiChrosorb SI 60 (5 μm) using a mobile phase of diethyl ether/methanol (70:30) by this HPLC method. Strychnine and brucine have also been separated by liquid–liquid partition chromatography on Corasil I coated with 1.1 per cent Poly G-300 using a mobile phase of heptane/ethanol (10:1) [300].

6.9.4   Cinchona alkaloids

Several cinchona alkaloids were separated on Corasil II coated with Poly G-300 using heptane/ethanol (10:1) as eluent [363]. A dynamic coating procedure was used to investigate the effect of stationary phase loading on the separations. When a 60 per cent saturated eluent was used no separation could be observed between quinine and cinchonine. Partial separation was obtained with a 40 per cent saturated eluent and the peaks were almost completely resolved when a 10 per cent saturated eluent was used. The separation of quinine, quinidine, cinchonine and cinchonidine was also investigated by Verpoorte and Svendson by adsorption chromatography [364].

6.9.5   Miscellaneous alkaloid separations

The dansyl derivatives of cephaeline, emetine, ephedrine and morphine have been analyzed by TLC and HPLC [390]. The improvement in selectivity and sensitivity obtained by using fluorimetric detection of these derivatives permits an analysis of these substances in the presence of ten- to a hundred-fold excesses of other drugs. Direct derivatization of syrups and aqueous slurries of capsules having a complex excipient and drug composition was feasible. HPLC was performed on silica gel columns (10 μm), using di-isopropyl ether/isopropanol/conc. ammonia (48:2:0.3) as the mobile phase for the syrup formulations and for the determination of emetine in capsules. For less polar components such as ephedrine, di-isopropyl ether saturated with conc. ammonia/isopropanol (99:1) was recommended. Detection limits were in the 1–10 ng range or better. The reproducibility of the method was limited by the derivatization step, but a coefficient of variation of less than 2 per cent could be obtained.

Several rauwolfia alkaloids (raupin, ajmaline, yohombine, reserpine and rescinnamine) have been separated by cation exchange chromatography [320]. The mobile phase was 0.02M ammonium hydrogen phosphate/methanol (70:30) adjusted to pH 7.0, and the separation was effected at 60° C. The alkaloids were detected at 254 nm.

6.10 Miscellaneous drugs

6.10.1 Hypoglycaemic agents

An HPLC procedure has been developed for the quantitative analysis of some sulphonylurea antidiabetic drugs (glyburide, chlorpropamide, tolazamide, tolbutamide and acetohexamide) in tablets [391]. The method used a column of Zipax HCP with either a methanol/borate buffer or methanol/citrate buffer as mobile phase. Testosterone, chlorpropamide or acetohexamide were used as internal standards for quantitation. The degradation products of the sulphonylureas did not interfere with the analysis of the parent drug. Precision was determined by assaying at least six samples of powdered tablets from single lots of each drug. The coefficient of variation ranged from 0.8 per cent for tolbutamide tablets to 2.56 per cent for tolazamide tablets. Another assay developed for the determination of chlorpropamide in tablet formulations involved adsorption chromatography on silica gel with a methanol mobile phase [392]. Bis(dodecyl)-phthalate was used as internal standard and the assay of four samples per day for four consecutive days gave a coefficient of variation of 2.7 per cent for the method.

Chlorpropamide or tolbutamide have also been determined in the presence of their metabolites in plasma [393]. The residue from an ether extract of acidified plasma was redissolved in the mobile phase, 0.05M ammonium formate/acetonitrile (83:17), and chromatographed on μBondapak C_{18}. UV detection was used and samples containing less than 0.5 μg ml^{-1} chlorpropamide or 5 μg ml^{-1} tolbutamide could be quantified. Tolbutamide has also been quantified in 1 ml serum samples by chromatography on Permaphase ETH, with a mobile phase of 0.01M sodium citrate/methanol (70:30) [394]. Using UV detection at 254 nm and with N-(p-methoxybenzenesulphonyl)- N'-cyclohexylurea as the internal standard, 2 μg tolbutamide could be determined.

6.10.2 Sympathomimetics

A series of seven imidazoline derivatives has been separated on Zipax SCX and SAX ion exchangers using an aqueous mobile phase buffered at pH 11.4–12.4 [395]. The extraction of the polar imidazoline bases from aqueous formulations into organic solvents is slow and direct analysis of the drug in the aqueous phase is advantageous. Using the HPLC method the intact drugs could be separated from their degradation products with minimal sample preparation. Also the separation of mixtures of the imidazolines in pharmaceutical preparations was possible.

6.10.3 Sulphinpyrazone

A method has been developed for the analysis of sulphinpyrazone in serum [396]. A slight modification of this approach allowed the simultaneous determination of sulphinpyrazone and its metabolites shown in Fig. 6.21, in plasma and urine [397]. In the original method 0.5 ml serum samples were acidified and extracted with 1-chlorobutane [396]. The extracts were chromatographed on a 250 x 2.2 mm i.d. column of MicroPak SI 10 using a mobile phase of dioxane/methanol (65:35). Using UV detection at 254 nm the minimum detectable concentration of sulphinpyrazone was 3 μg ml^{-1}. Subsequently, the drug and its metabolites were extracted from the acidified aqueous phase with 1-chlorobutane/ ethylene dichloride (80:20) [397]. Separation was obtained on a 120 x 4.7 mm i.d. column of LiChrosorb SI 60 (5 μm) using a mobile phase of dichloromethane/ethanol/water/acetic acid (79.1:19:1.9:0.002). The limit of detection was 0.2 μg ml^{-1}, using a 1 ml sample.

6.10.4 Sodium iodohippurate

Sodium iodohippurate, radioactively labelled with I^{131}, is used as a diagnostic aid in renal function determination. Problems have been experienced in its analysis; on paper chromatography the radioactive and non radioactive materials behaved differently [398], stable volatile derivatives were not obtained for GLC analysis, and low recovery from TLC plates led to poor precision. An HPLC method was developed which was independent of the amount of radioactive material present, thus permitting the determination of total o-iodohippurate [399]. The analysis was performed on SIL-X adsorbent with a mobile phase of chloroform/acetic acid (99:1) and a flow rate of 0.8 ml min^{-1}. Methyl-3,5-

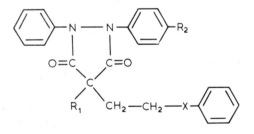

	R$_1$	R$_2$	X
Sulphinpyrazone	H	H	SO
Metabolites:			
I	OH	H	SO
II	H	H	SO$_2$
III	H	OH	SO

Fig. 6.21 Structure of sulphinpyrazone and its main human metabolites.

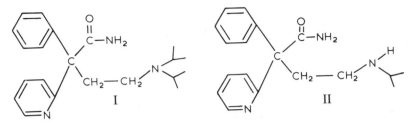

Fig. 6.22 Structure of disopyramide (I) and its mono-*N*-dealkylated metabolite (II).

dihydroxy benzoate was used as the internal standard and detection was at 254 nm. The major degradation product, *o*-iodobenzoic acid, eluted with the solvent front. Recoveries of added sodium *o*-iodohippurate ranged from 95.6—100.0 per cent. The accuracy and precision of the method was determined by 10 replicate injections of a solution of the compound by three analysts over four days. The mean value found was 98.8 per cent of the theoretical with a coefficient of variation of 1.6 per cent.

6.10.5   Cardiovascular drugs
Pentaerythritol tetranitrate, used in the treatment of anginal attacks, has been analyzed at the $2 \mu g \, ml^{-1}$ level in 5 ml blood by conversion to its tetra-*p*-methoxybenzoate derivative [400]. After extraction, chromatography was performed on a 600 × 2 mm i.d. column of Corasil II with heptane/chloroform (60:40) as mobile phase.

The antiarrhythmic agent disopyramide I (shown in Fig. 6.22), and its mono-*N*-dealkylated metabolite (II) were determined in plasma and urine using ion pair chromatography [401]. Extracts were chromatographed on a column of ODS-silica using methanol/water (53:47) containing heptanesulphonic acid as the mobile phase. The method was suitable for quantitation in the range 1—40 μg in urine and 0.05—4.0 μg in plasma, in samples of 0.1—1.0 ml.

The rapid and simultaneous determination of plasma levels of *N*-acetyl-procainamide and procainamide has been obtained by HPLC of extracts containing *N*-formylprocainamide as an internal standard [402]. The samples were separated on a column of μBondapak C_{18} using 0.4 per cent sodium acetate/acetic acid/acetonitrile (100:4:5) and detected at 254 nm.

6.10.6   Anaesthetics
A method which is suitable for the routine analysis of procaine in pharmaceuticals, has been developed [403]. Sample preparation was minimal and involved dissolution of the pharmaceutical product in the appropriate volume of water plus addition of pyrrocaine hydrochloride as internal standard. The samples were chromatographed on a column of μBondapak C_{18} using acetonitrile/water (60:40) containing 0.01 per cent concentrated ammonia, and

UV detection was used. Good peak shapes were obtained in the presence of ammonia and 0.25 μg of the free base could be quantified. The method was applied to the analysis of four different dosage forms and results comparable to those obtained by colourimetric and titrimetric methods were obtained.

6.10.7   Pilocarpine

Pilocarpine, effective in the treatment of glaucoma, contains an inactive stereoisomer. The separation of these isomers has been achieved on a 100 x 6 mm ion exchange column using 0.2M Tris buffer/isopropanol (95:5), at pH 9 [404]. The peaks were detected at 217 nm. The isomers could be assayed in the presence of one another and in the presence of excipients commonly found in commercial solutions. The limit of detection was 0.1 μg isopilocarpine in the presence of 100 μg pilocarpine.

Part III

Application of HPLC in biochemical analysis

7

Lipids

7.1 Introduction

The development of HPLC in the field of lipid analysis has been slower than in other areas because of the problems involved in detecting non-UV absorbing compounds. The RI detector, although frequently used, lacks sensitivity and cannot be used with the solvent gradient systems which are often needed to effect separation of non-polar and polar lipids. The most widely applicable detectors which have been developed are the transport flame ionization detectors (FIDs) [405], but several other systems such as the spray impact detector [82] and electrochemical detectors [406] have been applied to lipid analysis. The preparation of UV absorbing derivatives followed by UV detection has also been frequently used [109 112, 107 109].

Two comprehensive reviews on lipid analysis by HPLC have appeared [410, 411].

7.2 Total lipid analysis

Separation of total lipid mixtures have in the past been carried out by TLC or by adsorption column chromatography under gravity flow [412]. However, these methods are relatively inefficient and can be greatly improved by modern techniques involving HPLC. The major problem in the separation of the lipid classes is due to the very large differences in their polarity ranging from non-polar hydrocarbons such as squalene, to the polar phospholipids. Thus gradient elution systems, often involving the use of more than two solvents, are necessary to perform a complete lipid analysis.

The separation of reference mixtures of glycerides, fatty acids and sterols on silica columns has been reported [413, 414]. Stolyhwo and Privett separated mixtures of non-polar and polar lipids on 1 m x 2.8 mm columns of Corasil II (37–55 μm) which had been treated overnight with ammonium hydroxide [414]. As shown in Fig. 7.1 a three-step gradient system can be used to separate components ranging from fatty acid esters and glycerides to phospholipids, in 3 h. This method was applied to the analysis of the lipid material extracted from rat red blood cells [414] soybean 'lecithin' [415] and immature

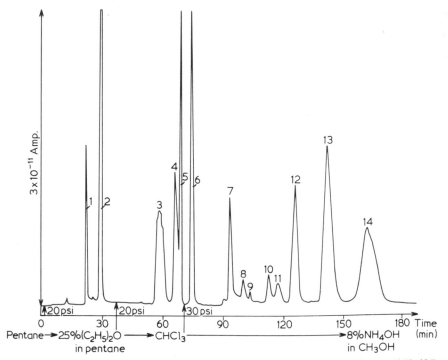

Fig. 7.1 Analysis of a total lipid mixture. Column packing, Corasil II (37—
55 μm); column dimensions, 1 m x 2 mm i.d.; eluent, three step gradient;
pentane to pentane/diethyl ether (75:25) to chloroform to methanol/
ammonium hydroxide (92:8); pressure, 20—30 psi (0.14—0.20 MPa); flow rate,
0.2 ml min⁻¹ column temperature, ambient; flame ionization detector. Solutes:
1, methyl oleate; 2, triolein; 3, cholesterol; 4, 1,3-diolein; 5, 1,2-diolein; 6,
monoolein; 7, 8, 9, 10, 11, components of beef brain cerebroside fraction
(10,11 = cerebrosides); 12, phosphatidyl ethanolamine; 13, phosphatidyl
choline; 14, beef brain sphingomyelin. Reproduced with permission from reference 414.

soybeans [416]. The lipids from soybean and soybean foods have also been
separated on a column of MicroPak SI-10 (10 μm) using gradient elution [413].
The column was initially eluted with hexane/chloroform (90:10) followed by an
ethanol concentration gradient of 3 per cent min⁻¹. Under these conditions the
triglycerides, diglycerides, sterols, free fatty acids and monoglycerides were
eluted in less than 15 min. Polar lipids such as phosphatidylserine, phosphatidylcholine and phosphatidylethanolamine had retention times of 18—23 min. The
minimum levels of detection, using an FID system, were 0.3—1.2 μg for
tristearin, distearin and monopalmitin and 2.8 μg for palmitic acid. Technical
epoxidized soybean oils have also been resolved into peaks for triglycerides
bearing increasing numbers of oxiran oxygen atoms per triglyceride molecule by
adsorption chromatography on silica using gradient elution techniques [405].
    The rapid separation of non-polar lipid classes has been achieved by exclusion
chromatography on columns of 200 Å or 60 Å Poragel using aqueous acetone

mobile phases [417] or by adsorption chromatography using gradient pro-
grammes [405].

Neutral serum lipids have been separated on a 250 x 2.2 mm column of silica
gel H using flow programming with an eluent of heptane/diethyl ether
(80:20) [418]. Phospholipids (mainly phosphatidylcholine and sphingomyelin)
were separated on a second silica column using chloroform/methanol/water
(65:25:4) as mobile phase.

HPLC has also been used for the quantitative determination of the mono-, di-,
and triglycerides in technical glycerol monostearate [419] and glyceride-based
lubricants [420] and for the analysis of the unchanged triglyceride content in
used frying oils [405].

7.3 Fatty acids

Fatty acids are the basic components of most naturally-occurring lipids and the
determination of fatty acid profiles is of considerable importance in lipid
analysis. A great deal of work in this field has been accomplished by GLC but
recently HPLC has been increasingly applied to the problem of fatty acid
analysis.

Free fatty acids have been separated on an ion exchange resin [406] or by
reversed phase chromatography on C_{18} bonded materials [82, 421]. Pei et al.
separated the free fatty acids ranging from hexanoic to dodecanoic acids
(6:0–12:0) and from arachidic to lignoceric acids (20:0–24:0) in a few min at
50° C on a 1 m x 2 mm i.d. column of Vydac reversed phase support
(35–44 μm) using mobile phases of methanol/water (60:40) and (90:10)
respectively [421]. Separations of the C_{18} and C_{20} polyunsaturated fatty acids
were also obtained. The separation of 5, 8, 11, 14, 17-eicosapentaenoic acid
(20:5); 5, 8, 11, 14-eicosatetraenoic acid (20:4); and 11, 14, 17-eicosatrienoic
acid (20:3) at 60° C in 5 min using a methanol/water (80:20) mobile phase is
shown in Fig. 7.2. This resolution could be of particular interest since these are
the precursors of prostaglandins E_3, E_2 and E_1 respectively. These separations
were all monitored with a refractive index detector, and 100–500 μg of each
component was chromatographed. However, as little as $0.5–2 \times 10^{-7}$ moles of
several short chain fatty acids, chromatographed on an ion exchange resin with
an eluent of water/2-methoxyethanol (90:10), could be detected using a flow
coulometric detector [406]. The electrolyte solution consisted of $0.01M$
p-benzoquinone/$0.001M$ hydroquinone/$0.1M$ potassium nitrate. In addition,
nanogram amounts of C_8, C_9 and C_{10} fatty acids were detected using the spray
impact detector after separation on a column of Bondapak C_{18}/Corasil with an
aqueous eluent [82].

Silver nitrate TLC is known to separate saturated compounds from some
unsaturated analogues, and compounds containing cis-double bonds from their
isomers containing trans-double bonds because of the differences in the
silver-complexation constants of the alkenes. Mikes et al., using this principle,
separated stearic (C_{18}-saturated) elaidic (C_{18}-trans unsaturated) and oleic

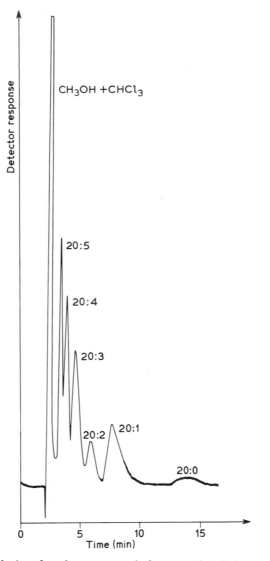

Fig. 7.2 Analysis of polyunsaturated fatty acids. Column packing, Vydac reversed phase support (35−44 μm); column dimensions, 1000 x 2 mm i.d.; eluent, methanol/water (80:20); pressure, 1000 psi (6.8 MPa); flow rate, 0.7 ml min^{-1}; column temperature, 60°C; refractive index detection. Solutes: 1, 5,8,11,14,17-eicosapentaenoic acid (20:5); 2, 5,8,11,14-eicosatetraenoic acid (20:4); 3, 11,14,17-eicosatrienoic acid (20:3); 4, eicosadienoic acid (20:2); 5, eicosenoic acid (20:1); 6, arachidic acid (20:0) (all around 100 μg). Reproduced with permission from reference 421.

(C_{18}-cis unsaturated) acids as their methyl esters on a column of Corasil II loaded with 0.8 per cent silver nitrate and 1.75 per cent ethylene glycol [422]. A mobile phase of hexane/heptane (50:50) was used and the acids were detected using a refractive index detector. The same separation was performed in 10 min on a 170 x 4 mm column of Porasil A coated with 25 per cent silver nitrate using a mobile phase of heptane/diisopropyl ether (80:20) [405]. In a variation of this, Schomburg used a reversed phase column of LiChrosorb RP8 (10 μm) in conjunction with an aqueous isopropanol mobile phase containing 0.5—2.5 per cent silver nitrate for the separation of olefins [423]. The olefins which formed the stronger metal complexes were preferentially partitioned into the mobile phase and hence eluted faster. The methyl esters of oleic and elaidic acid were separated in 20 min using this system, the elution order being the reverse of that obtained by the previous method.

Chain length and degree of unsaturation form the basis of the reversed phase separation of fatty acid methyl esters [421, 424]. The methyl esters of the fatty acids from linseed oil, viz., methyl linolenate, methyl linoleate, methyl stearate and methyl palmitate were chromatographed on Bondapak C_{18}/Corasil with acetonitrile/water (80:20) as eluent and RI detection [424]. Good separations were also obtained for the esters of the fatty acids from soybean, safflower, corn and olive oil. The same method was carried out on a preparative scale using two 600 x 8 mm i.d. stainless steel columns packed with Bondapak C_{18}/Porasil [425]. Up to 200 mg samples of pure esters or fractions from hydrogenated fats could be collected for further investigation. 10—400 μl of sample solution were injected by syringe through a septum at operating pressures of 300—500 psi (2.0—3.4 MPa), corresponding to flow rates of 5—10 ml min^{-1}.

The detection limits for fatty acids have been improved by the preparation of suitable UV absorbing derivatives. Benzyl esters of palmitic, heptadecanoic and stearic acids were prepared using 1-benzyl-3-p-tolyltriazene as derivatizing agent, and the derivatives were separated on Corasil II with chloroform/heptane (50:50) as mobile phase [401]. The lower limit of detection of benzyl stearate was 1—2 μg. Preparation of the p-nitrobenzyl esters of lauric, myristic, palmitic and stearic acids before chromatography, allowed detection in the picomole range [426].

Naphthacyl esters of C_{18} and C_{20} fatty acids were separated by reversed phase chromatography on a 900 x 1.75 mm i.d. column of C_{18}/Corasil with methanol/water (85:15) as eluent [110]. Lower limits of 4 ng of the esters were detected. In the C_{18} series, α- and γ-linolenic acid esters were separated from linoleic and oleic derivatives, and in the C_{20} series the esters of arachidonic and dihomo-γ-linolenic acids were resolved. Phenacyl esters of C_{12}—C_{24} fatty acids have been separated by gradient elution on μBondapak C_{18} using acetonitrile/water eluents [109] as shown in Fig. 7.3. The lower limit of detection was ~100 ng for the C_{18} acid at 254 nm. Quantitative formation of p-bromophenacyl esters was achieved using crown ether catalysts (19-crown-6 or dicyclohexyl-18-crown-6) [111]. Since almost any solvent could be used for the

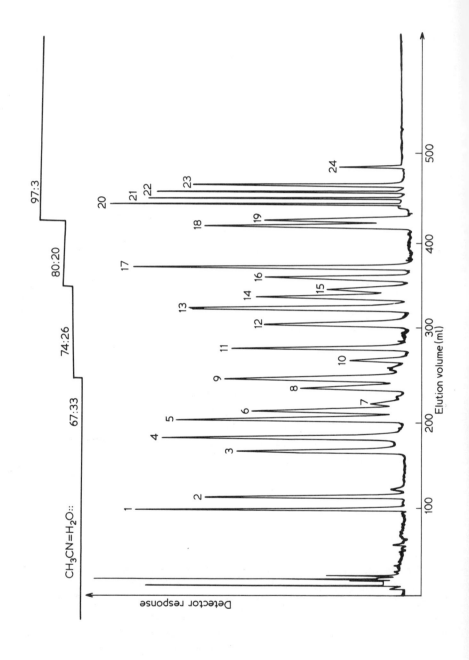

reaction and rigorously anhydrous conditions were not necessary, the reaction mixtures could be directly injected into the chromatograph. The separation of the derivatives of the C_2-C_{20} acids was achieved at $40°C$ on a column of C_9/Corasil II using methanol/water eluents. Between 0.25 and 1 μg of each acid derivative was injected and the detection limits were 1 ng of C_2 acid and 50 ng of C_{20} acid at 254 nm.

Linoleic acid is oxygenated by potato lipoxygenase to give mainly 9-hydroperoxy-10,12-octadecadienoic and 13-hydroperoxy-9,11-octadecadienoic acids. The hydroperoxides had previously been considered too unstable for analysis, but their methyl esters were separated on Partisil-10 silica, giving ratios of 9:13-hydroperoxides of 95:5 and 6:94, using potato and soybean lipoxygenases respectively [427].

A new assay method, using HPLC, has been developed which permits the simultaneous isolation, determination and quantitation of lauric acid and its hydroxylated metabolites as their methyl esters, after incubation with kidney or liver microsomes [428]. Methyl 11-hydroxylaurate and methyl 12-hydroxylaurate were separated on a μBondapak C_{18} column using a mobile phase of methanol/water (60:40). The strongly retained methyl laurate could then be eluted with 100 per cent methanol. The peaks were detected with a refractometer but quantitation was performed with radioactive samples which were collected and measured by scintillation counting.

7.4 Phospholipids

The separation and quantitative analysis of phospholipids containing primary amine groups has been achieved by chromatographing their biphenylcarbonyl derivatives [409]. The phospholipid derivatives were separated on a 200 x 2.1 mm column of MicroPak SI-10 silica, using one of two mobile phases. For phosphatidylethanolamine derivatives, dichloromethane/methanol/15M ammonia (92:8:1) was used and for phosphatidylserine and lysophosphatidylethanolamine derivatives the same solvents were used in the proportions (80:15:3). The response for the phosphatidylethanolamine and phosphatidyl-

---

Fig. 7.3 Analysis of fatty acid phenacyl esters. Column packing, μBondapak C_{18} (10 μm); column dimensions, 900 x 6.4 mm i.d.; eluent, stepped gradient from acetonitrile/water (67:33) to (97:3), flow rate, 2.0 ml min^{-1}; column temperature, ambient; detection, UV at 254 nm. Solutes: phenacylesters of 1, lauric (12:0); 2, myristoleic (14:1); 3, α- and γ-linolenic (18:3); 4, myristic (14:0); 5, palmitoleic (16:1); 6, arachidonic (20:4); 7, transpalmitoleic (trans 16:1); 8, linoleic (18:2); 9, pentadecanoic (15:0); 10, linolelaidic (trans 18:2); 11, eicosatrienoic (20:3); 12, palmitic (16:0); 13, oleic (18:1, Δ^9) and vaccenic (18:1, Δ^{11}); 14, petroselinic (18:1, Δ^6); 15, elaidic (trans 18:1); 16, eicosadienoic (20:2, $\Delta^{11,14}$); 17, heptadecanoic (17:0); 18, stearic (18:0); 19, eicosaenoic (20:1, Δ^{11}); 20, nonadecanoic (19:0); 21, arachidic (20:0) and erucic (22:1); 22, heneicosanoic (21:0); 23, behenic (22:0) and nervonic (24:1); 24, lignoceric (24:0) acids. Reproduced with permission from reference 109.

serine derivatives at 268 nm was linear over the range of 10 pmole to 20 nmole injected. The coefficient of variation for 1 nmol injections of the same sample was 0.5 per cent and the lower limit of detection of the phospholipid derivatives was around 10–13 pmole (0.3–0.4 ng of phosphorus). The quantitative analysis of aminophospholipids in rat brain and liver extracts, as well as rat brain myelin and microsomal fractions, gave results which agreed well with TLC and literature values.

A sensitive method for the separation of phosphatidylcholine and sphingomyelin by chromatography on a microparticulate silica column using acetonitrile/methanol/water (75:21:4) as solvent has been developed [429, 430]. The elution of the phospholipids was monitored with a UV spectrophotometer at 203 nm. It was shown by using synthetic phosphatidylcholines of known fatty acid composition and of varying degrees of unsaturation that the absorption at 203 nm was primarily due to the isolated double bonds and the response measured depended on the number of double bonds in the molecule. Approximately 1 nmole of phosphatidylcholine, containing at least one double bond per molecule, could be detected. Quantitation of phosphatidylcholine and sphingomyelin could be achieved using UV absorption, or alternatively, the peaks were collected and the phospholipids were determined by phosphorus analysis. The analysis of phosphatidylcholine and sphingomyelin present in the lipid extracts from animal tissues, blood and amniotic fluids was made without interference from other phospholipids or UV absorbing material.

In order to study the degradation of lecithins in septic bile ducts a method was developed for the separation and analysis of glycerophosphorylcholine, phosphorylcholine, glycerophosphate and orthophosphate [431]. The separation was achieved by gradient elution on a 150 x 6 mm o.d. column of Dowex 1–X4 (Cl⁻) anion exchange resin and the components were detected by automatic phosphorus analysis of collected fractions.

7.5 Ceramides

Ceramides (*N*-acylsphingosines) constitute the basic lipid moiety of sphingolipids.

Excess amounts of hydroxy and nonhydroxy fatty acid ceramides (HFA- and NFA-ceramides respectively) have been reported in tissues of patients suffering from Farber's disease, a lipid storage disease. Recent studies have shown that HPLC analysis of urinary ceramides may be the simplest procedure for detecting the metabolic disturbance in this disease. In one study *O*-benzoylated HFA- and NFA-ceramides were chromatographed on 27 μm Zipax with 0.05 per cent methanol in pentane as eluent [432]. Using labelled representatives of both types of ceramide, it was shown that the overall recoveries of derivatized ceramide through the entire procedure of extraction (from tissue homogenates or urine), hydrolysis, column chromatography and benzoylation was 80 per cent. NFA- and HFA-ceramides from tissue and body fluid extracts were analyzed in 3 min. In a series of 12 samples of blood sera from control patients, the total ceramides were measured as 4.70 μg ml^{-1}. Only trace amounts of HFA-ceramides were observed. Analysis of sera from patients suffering from Farber's disease gave values of 9.5 μg ml^{-1} of NFA-ceramides and 4.0 μg ml^{-1} of HFA-ceramides. These values were higher than the control levels, but the difference was not as great as that observed in some tissues and urine. Farber's disease urine contained at least 200 times as as much ceramide per mg of creatinine as the controls. The measured levels in urine from diseased patients were 1.0 μg NFA-ceramide and 0.2 μg HFA-ceramide, but no ceramide was demonstrated in 28 control urine samples with comparable creatinine content. The value of 1.2 μg ceramide per mg creatinine in Farber's disease urine was confirmed in an independent HPLC study of ceramide analysis [433]. In this case, benzoylated ceramides were chromatographed on MicroPak SI-10 silica with 0.025 per cent methanol in pentane as the mobile phase, and fractions corresponding to each peak were collected. The minimum detectable amount of ceramide was 6 ng per mg of creatinine although examination at this level involved a lengthy pre-analysis clean-up of the extracts.

7.6   Glycolipids

Evans and McCluer have devised a system for the separation of neutral glycosphingolipids [408] and cerebrosides containing hydroxy and non-hydroxy fatty acid residues [434] as their benzoylated derivatives. Mono-, di-, tri- and tetraglucosyl- ceramides from serum were resolved in 25 min at 40°C on a 800 x 2.1 mm i.d. column of Zipax using gradient elution with methanol in hexane, and detected at 254 nm [408].

8
Steroids

HPLC has been successfully applied to the separation of all classes of steroid hormones [107, 435–437]. The technique has the advantage that thermally labile steroids, for example corticosteroids, and involatile samples such as steroid conjugates [107, 438] can be analyzed relatively easily. A major problem has been one of low detector sensitivity but this can be overcome by the preparation of UV absorbing derivatives [106-108]. (Comparative normal and reversed phase separations of a steroid mixture were shown in Figs. 5.2 and 5.3.)

8.1 Corticosteroids

One of the earliest applications of HPLC to corticosteroid analysis involved liquid–liquid partition chromatography between a water or water/methanol mobile phase and a stationary phase consisting of Amberlite LA-1 [n-dodecenal (trialkylmethylamine)] coated on a polytrifluoroethylene support [436]. Fourteen corticosteroids could be separated on this column using flow programming. Standard mixtures of corticosteroids have also been resolved by liquid–liquid partition chromatography on columns of BOP/Zipax using a mobile phase of heptane/tetrahydrofuran (80:20) [107] or by partition chromatography on silica columns using ternary solvent systems similar to those proposed by Huber et al. [439], consisting of dichloromethane/ethanol/water [440–443], dichloromethane/methanol/water [444] or isooctane/ethanol/water [445].

HPLC has also been applied to the determination of corticosteroids in human plasma [440, 442, 445–448]. Both adsorption chromatography on SIL-X, with a mobile phase of chloroform/dioxan (20:1), and reversed phase chromatography on an ODS-bonded silica at 40°C with water/methanol (60:40) as eluent, were investigated [447]. Although cortisol, cortisone and aldosterone eluted as one peak in the latter system this method was chosen for the determination of plasma corticosteroids [448]. Since extraction, chromatography and quantitation were fast, it was possible to perform 20–30 determinations per day. Subsequently methods for the routine determination of cortisol levels in human plasma were devised [440, 445]. Both methods were based on the extraction of cortisol from plasma with dichloromethane, followed by chromatography on silica using ternary solvent systems. Meijers *et al.* could determine 8.5 ng of cortisol with a coefficient of variation of 20 per cent and 100 ng with a coefficient of variation of 3.3 per cent at 240 nm [445]. Using prednisolone as an internal standard, and monitoring simultaneously at 230 nm and 254 nm, Trefz *et al.* obtained a limit of detection for cortisol of 10 ng per ml using 1 ml of plasma with a coefficient of variation of 5 per cent [440].

Cortisone and cortisol (hydrocortisone) are very potent anti-inflammatory agents which are used extensively in the treatment of rheumatoid arthritis, and acute inflammation of various tissues. Many synthetic analogues of these corticosteroids have been produced and are constituents of pharmaceutical preparations. Assays based on HPLC, have been reported for hydrocortisone [236, 250], fluocinolone acetonide [235], triamcinolone acetonide [340] and dexamethasone disodium phosphate [107] in tablets and other formulations. These assays have been discussed in Section 6.6.3. The determination of methyl prednisolone, used in the treatment of bovine mastitis, in milk from treated cows, has also been achieved by HPLC [449], as has the determination of synthetic corticosteroids in plasma [446, 450].

The microbiological transformation of steroids by the introduction of an 11β-OH group is becoming increasingly important. The control and assessment of material produced during fermentation was achieved, in the case of the corticosteroid shown in Fig. 8.1, by direct analysis of the fermentation mixture on Permaphase ODS at 40°C using a mobile phase of water/methanol (92.5:7.5) [235].

O'Hare *et al.* used reversed phase HPLC with gradient elution on Zorbax ODS columns to separate, identify and measure spectrophotometrically the steroids secreted by both human adrenal and testicular cells in primary monolayer culture [435]. The separation of all the steroids could not be achieved within a reasonable period of time (<1 h) using a single system. Three separate systems were therefore developed; a methanol/water gradient was used for the adrenal steroids; an acetonitrile/water gradient for the testis steroids; and the polar steroids such as aldosterone and 6β-hydroxycortisol were resolved using a dioxan/water gradient as shown in Fig. 8.2. These three systems together permitted the resolution of at least 43 naturally-occurring steroids, plus four synthetic steroids with adrenocortical activity. The minimum detectable

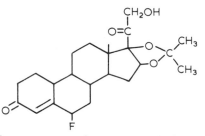

Fig. 8.1   Structure of a representative corticosteroid.

quantity of steroid, obtained by monitoring at 240 nm for the Δ-4-en-3-one chromophore (254 nm for dioxan solvents) was 2 ng, measured as twice baseline noise. In practice, culture medium samples containing >5 ng (well below the levels encountered in most experiments) could be easily distinguished from the background.

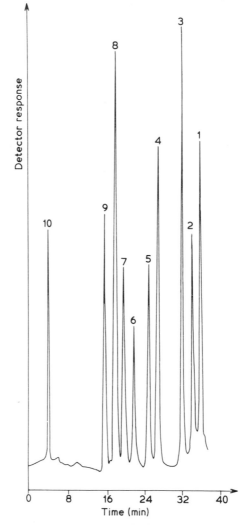

Fig. 8.2 Analysis of polar adrenal steroid standards. Column packing, Zorbax ODS; column dimensions, 250 x 2.1 mm i.d.; eluent, concave gradient from water/dioxan (80:20) to dioxan; column temperature, 45°C; detection, UV at 254 nm; sensitivity, 0.16 AUFS. Solutes 1, corticosterone; 2, 18-hydroxy-11-deoxycorticosterone; 3, 11β-hydroxyandrostenedione; 4, cortisol; 5, cortisone; 6, 18-hydroxycorticosterone; 7, aldosterone; 8, 18-hydroxy-11-dehydro-corticosterone; 9, 11-dehydroaldosterone; 10, 6β-hydroxycortisol. Reproduced with permission from reference 435.

8.2 Oestrogens

Oestrogen analysis in female urine is important for the monitoring of pregnancy and for the examination of non-pregnant females who are taking fertility drugs. During pregnancy large amounts of oestrogens especially oestriol ($\sim$3 μg ml^{-1}) are excreted. Deficiencies in the excreted levels are indicative of possible malfunction of the placenta. It has been suggested that the incidence of multiple births among females who are taking fertility drugs could be associated with a high concentration of oestriol in the body at the time of conception. However, it is more difficult to determine the oestrogens in the non-pregnancy case since the normal level of each oestrogen is around 3 ng ml^{-1}.

Several methods have been reported for the analysis of urinary oestrogens by HPLC [437, 451–453]. In one procedure the extracted oestrogen conjugates were hydrolyzed, chromatographed on Corasil I using ethanol/hexane as eluent [451] and monitored at 280 nm. However, the oestrogens eluted on the tail of an impurity peak and it was necessary to use a tangent method for quantitation. The levels of oestradiol and oestriol found in pregnacy urine were 0.6 μg ml^{-1} and 1.93 μg ml^{-1} respectively. Oestriol was not detected in non-pregnancy urine. Under the same chromatographic conditions, the minimum detectable amounts of oestradiol, oestriol and oestrone were 13 ng, 19 ng and 23 ng respectively.

This assay procedure has been considerably improved by the use of microparticulate packing materials [452]. The increased capacity of these materials allowed the use of larger samples without significant loss in efficiency and hence increased the accuracy of the determinations. For the determination of oestriol, an adsorption method using Partisil-5 and a mobile phase of hexane/ethanol (95:5), was found to be adequate. However, for the quantitative determination of oestriol, oestradiol and oestrone, reversed phase chromatography on Partisil-10 ODS was recommended. As shown in Fig. 8.3. complete resolution of the three oestrogens was obtained with a mobile phase of methanol/0.1 per cent aqueous ammonium carbonate (55:45).

In another assay the quantitative determination of oestriol in pregnancy urine was achieved in 20 min by liquid–liquid partition chromatography [437]. A concentration of $\sim$3 μg ml^{-1} of oestriol could be determined with a coefficient of variation of 5 per cent.

The resolution of a mixture of oestrogens, namely oestrone, oestradiol, oestriol-3-methylether, 16α-hydroxyoestrone, 16-keto-oestradiol, oestriol, 17-epioestriol and 16-epioestriol was examined on several supports including μBondapak NH$_2$, μPorasil and μBondapak C$_{18}$ [453]. The best separations were obtained on a 300 x 4 mm column of μBondapak NH$_2$ using hexane/isopropanol (80:20) as eluent and this method was applied to the examination of oestrogenic hormones in pregnancy urine.

The separation of nine equine oestrogens (oestrone, equilin, equilenin, 17β-oestradiol, 17α-oestradiol, 17β-dihydroequilin, 17α-dihydroequilin, 17β-dihydroequilenin and 17α-dihydroequilenin) was reported by both liquid–liquid

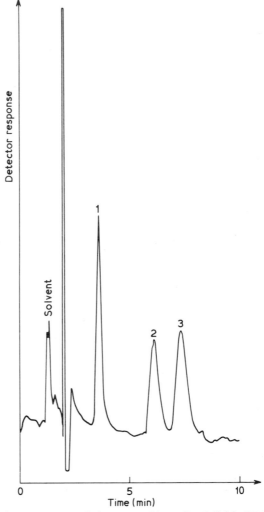

Fig. 8.3 Analysis of oestrogens. Column packing, Partisil-10 ODS; column dimensions, 250 x 4.6 mm i.d.; eluent, methanol/0.1 per cent aqueous ammonium carbonate (55:45); flow rate, 2 ml min^{-1}; column temperature, ambient; detection, UV at 280 nm; sensitivity, 0.25 AUFS. Solutes: 1, oestriol; 2, oestrone; 3, oestradiol. Reproduced with permission from reference 452.

partition chromatography on BOP/Zipax [454] and by chromatography on Permaphase ETH[454, 455]. The analysis of all nine oestrogens was accomplished on Permaphase ETH in 45 min using a mobile phase of hexane/tetrahydrofuran (98:2) at 40°C [454].

Oestradiol has been analyzed quantitatively in tablets by chromatography on Permaphase ODS using a mobile phase of water/methanol (80:20) with toluene as the internal standard [235]. Alternatively, the purity of oestradiol samples has been assessed using a BOP/Zipax system with a mobile phase of heptane/tetrahydrofuran (95:5) [107]. The separation of oestradiol glucosiduronic acid, a naturally occurring urinary steriod conjugate, from several impurities was achieved on a strong anion exchange column (Zipax SAX) using an ionic strength gradient [107].

Several synthetic oestrogens from pharmaceutical preparations were chromatographed on a 1 m x 2.1 mm i.d. column of Permaphase ETH at 60°C using water/isopropanol (75:25) as mobile phase [456]. Using a UV monitor set at 254 nm the limit of detection for diethylstilboestrol was 1 ng (signal: noise = 2:1). The pharmaceutical preparations were extracted with combinations of ether, dimethylformamide and methanol. Quantitation was achieved by measurement of the peak area of the sample relative to an internal standard (phenothiazine). Drug stability studies were also performed in which diethylstilboestrol was photolytically decomposed and then separated from its three decomposition products by chromatography. The separation of diethylstilboestrol from methyl testosterone and reserpine, two drugs with which it is often found in combination, was also achieved in the same system.

8.3 Androgens and progestagens

8.3.1 Separation of free steroids
Standard mixtures of androgens and progestagens have been resolved by liquid–liquid partition chromatography [107, 457] and reversed phase chromatography [107, 438]. Progesterone, androsterone, testosterone, 19-nortestosterone and 11-ketoprogesterone were resolved on a BOP/Zipax system using a heptane mobile phase [107]. Similar mixtures were separated on silica columns coated with formamide using heptane/dioxane (95:5) as eluent [457] or on Permaphase ODS by gradient elution with aqueous methanol [107]. Both refractive index and UV detection were used.

The separation of four free 17-ketosteroids viz., androsterone, epiandrosterone, etiocholanolone and dehydroepiandrosterone, along with their sulphate and glucuronide conjugates, was achieved by reversed phase chromatography on a MicroPak CH column [438]. The best mobile phase for the separation of the free steroids was found to be water/methanol (50:50), while 80 per cent and 70 per cent water gave optimum resolution of the sulphates and glucuronides respectively. The solutes were monitored using a refractive index detector.

An HPLC method has been developed for the simultaneous identification and quantitation of the anabolic steroid methandrostenolone and its impurities methyltestosterone, 6α,17β-dihydroxy-17α-methylandrosta-1,4-dien-3-one and 6β,17β-dihydroxy-17α-methylandrosta-1,4-dien-3-one [458]. Separation was effected on a 250 x 2.1 mm i.d. column of LiChrosorb SI 60 using a mobile phase of hexane/isopropanol/dichloromethane (82:15:3); m-dinitrobenzene was used as the internal standard for quantitation and detection was performed at 254 nm.

A rapid method for the quantitation of 17α-ethynyl-17β-acetoxy-19-norandrost-4-en-3-one oxime in tablets involved chromatography on a column of Permaphase ODS at 60° C with a methanol/water (50:50) mobile phase (459).

Slocum and Studebaker have developed an assay of the bacterial enzymes which catalyze the introduction of 1,2 double bonds in steroids by chromatography of acidified samples of the reaction solutions [460]. The following reactions were examined:

androst-4-ene-3,17-dione $\longrightarrow$ androst-1,4-diene-3,17-dione

progesterone $\longrightarrow$ Δ^1-progesterone

Reichstein's compound S $\longrightarrow$ Δ^1-Reichstein's compound S

As the steroids in these reactions all absorb strongly at 254 nm, UV monitoring was used to detect both substrate and product in the column effluent. The steroids were separated by reversed phase chromatography on a μBondapak C_{18} column using water/acetonitrile eluents. The preparative scale chromatography of cholesteryl phenylacetate [181, 183] and progesterone [181] has been mentioned in Section 4.9.

8.3.2 Separation of derivatives
For steroids which have no appreciable UV chromophore, problems of detection must be overcome. Refractive index detection is, of course, possible but with a serious loss in sensitivity (refractive index monitors are usually 2–3 orders of magnitude less sensitive than UV detectors). The normal solution to the problem lies in the formation of steroid derivatives which have strong UV chromophores. Since many steroids have a hydroxyl or a ketone functional group, formation of benzoate or nitrobenzoate esters of hydroxyl groups and 2,4-DNP derivatives of ketones are feasible processes.

(a) Non UV absorbing hydroxy steroids
The benzoate esters of Δ^5-pregnenolone, a primary intermediate in steroid metabolism, and allopregnanolone, a urinary metabolite of progesterone, were separated on a 1 m column of Permaphase ODS using methanol/water (66:34) as mobile phase [108]. Allopregnanolone benzoate was also separated from the benzoate derivative of 5α-pregnan-3β,20β-diol on the same column using methanol/water (85:15) as mobile phase. Similarly, the benzoates or p-nitrobenzoates of androsterone, epiandrosterone, dehydroepiandrosterone, Δ^5-pregnanolone and allopregnanolone could be resolved in the former system. For benzoates (λ_{max} = 230 nm) with $k' < 2$, about 10 ng could be detected at 254 nm. An increase in sensitivity resulted when the p-nitrobenzoate esters were used, as these have an absorbance maximum at 254 nm and 1 ng of these esters could be detected [108].

(b) Non UV absorbing 17-ketosteroids
The analysis of non UV absorbing 17-ketosteroids by HPLC has been accomplished by prior formation of 2,4-DNP derivatives [106, 107]. Neutral

17-ketosteroids are of clinical interest in the diagnosis of certain illnesses. In normal males, about 70 per cent of the total urinary 17-ketosteroids is derived from the adrenal cortical hormones, the remainder arising from the metabolism of testosterone. In the female almost all the 17-ketosteroids are derived from the adrenal cortical hormones. Thus monitoring of 17-ketosteroid levels in the male is indicative of adrenal and testicular function, and in the female of adrenal function alone. The major 17-ketosteroid produced by the adrenal gland is dehydroepiandrosterone sulphate, and the principal metabolites, appearing as conjugates in the urine, are androsterone, epiandrosterone and etiocholanolone. The 11β-hydroxyl and 11-keto derivatives of androsterone and etiocholanolone may also be found in urine.

Liquid–liquid partition chromatography of the 2,4-DNP derivatives from an acid hydrolysate of male urine on a 1 m column of BOP/Zipax with an isooctane mobile phase, showed the presence of androsterone, epiandrosterone, etiocholanolone and dehydroepiandrosterone [106]. On BOP/Zipax columns, the more polar, physiologically important 11β-hydroxylated derivatives of androsterone and etiocholanolone were excessively retained, but these were separated by reversed phase chromatography on a 1.2 m x 1.8 mm i.d. column of C_{18}/Corasil using ethanol/water (50:50) as eluent [106]. After selective extraction of the sulphate conjugates of 17-ketosteroids from urine followed by hydrolysis, derivatization and analysis, the relative amounts of epiandrosterone and dehydroepiandrosterone were found to be greater than in extracts of free steroids, confirming that these steroids are predominantly excreted as sulphate conjugates [106]. The 17-ketosteroids from a plasma extract were also derivatized and chromatographed using this method and androsterone and dehydroepiandrosterone were identified by comparison of their retention times with those of standards [106]. About 1 ng of the 2,4-DNP derivative of androsterone could be detected at 254 nm compared with 1 μg of the underivatized steroid [107]. Also the 2,4-DNP derivatives of androsterone and epiandrosterone were separated on a BOP/Zipax column using heptane as eluent, whereas the parent steroids could not be resolved under the same conditions [107]. Thus derivatization can also alter selectivity, and an increase in resolution may be achieved by derivative formation.

8.4 Phytosterols

The separation of the acetates of C_{27}, C_{28} and C_{29} sterols and of sterols differing in degree of unsaturation has been accomplished by reversed phase chromatography on μBondapak C_{18} using a mobile phase of methanol/chloroform/water (71:16:13) [461]. The lower limit of detection of steryl acetates using the RI detector was approximately 50 μg. The sensitivity of detection could be greatly increased by using the steryl benzoates which could be monitored by UV, but the resolution in this case was inferior. The HPLC method was not as selective as silver nitrate–silica gel TLC in distinguishing the position of unsaturation in the sterol molecule.

8.5 Sapogenins and cardiac glycosides

The benzoate derivatives of the major sapogenins (hecogenin, $\Delta^{9(11)}$-dehydrohecogenin and tigogenin) occurring in the *Agave* species have been analyzed by reversed phase HPLC on a 250 x 4 mm i.d. column of LiChrosorb RP8 using an acetonitrile/water (80:20) mobile phase [462]. Sarsasapogenin, $\Delta^{9(11)}$-dehydrotigogenin, and diosgenin were also separated in the same system.

The cardiac glycosides are a therapeutically important group of plant constituents. Several digitalis glycosides have been separated by ion exchange [463], reversed phase [107, 464] or adsorption [465] chromatographic methods. Digoxin and its metabolites (digoxigenin, digoxigenin mono-digitoxoside and digoxigenin bisdigitoxoside) could be separated isocratically on μBondapak C_{18} using water/acetonitrile (75:25) [464]. Similarly digitoxin and its metabolites could be resolved on the μBondapak C_{18} column using water/acetonitrile (67:33) [464], and on Zipax SCX with a mobile phase of water/n-pentanol (96:4) at 45°C [463]. However, simultaneous resolution of all eight compounds required gradient elution on μBondapak C_{18} [464]. The minimum detectable amounts of these compounds depend on the mode of separation (isocratic or gradient elution) as well as the molecular weight of the the compound. Using detection at 220 nm (λ_{max} for the butenolide ring) as little as 4–20 ng could be quantitated [464]. By using two solvent systems [isooctane/n-pentanol/acetonitrile/water (124:35:12:2) for low polarity glycosides and heptane/t-butanol/acetonitrile/water (712:204:93:10.4) for glycosides of medium and high polarity], it was possible to separate 14 digitalis glycosides on LiChrosorb SI-60 [465]. Detection limits as low as 15 ng for digitoxigen and 25 ng for digitoxin (in 5 μl injections) were obtained by monitoring at 220 nm.

These methods of analysis should be applicable to the assay of clinical preparations of digitalis glycosides. However, the application to clinical studies in body fluids is less obvious. The usual therapeutic range for serum digoxin concentrations is 0.7–1.5 ng ml^{-1}, which is beyond the sensitivity of these methods at present.

Improved sensitivity was achieved in the analysis of cardiac glycosides by derivatization with p-nitrobenzyl chloride [466, 467]. Not only did the derivatization enhance the UV absorbance of the molecules at 254 nm, it also rendered them less polar, resulting in better chromatographic properties. Rapid isocratic separations with reduced tailing of complex glycoside mixtures were obtained on silica with hexane/dichloromethane/acetonitrile (10:3:3) as the mobile phase. The detection limits varied between 5 and 20 ng ml^{-1} at an injection volume of 100 μl.

8.6 Steroidal alkaloids

The steroidal alkaloids, tomatidine, solanidine, solasodine, rubijervine, veratramine and jervine have been separated on a 2.4 m column of Porasil A using a hexane/acetone gradient elution system [468]. The method was used to isolate crystalline solanidine from a crude mixture of aglycones obtained from

Solanum chacoense, and to separate radioactive solanidine from extracts of potato plants fed with $[4-^{14}C]$-cholesterol.

8.7 Miscellaneous steroids

8.7.1 Saturated steroid hydrocarbons

The separation of a range of isomeric and homologous steranes, including epimeric pairs of 24-alkyl-5α-cholestanes has been achieved by adsorption chromatography on active alumina [469]. These results were obtained on a low resolution system and it should be possible to improve the separations using microparticulate alumina.

8.7.2 Bile acids

A partial separation of the major conjugated bile acids of human bile has been achieved by chromatography on Corasil II or μPorasil [470]. Initially a group separation was obtained using the alkaline solvent system, ethyl acetate isopropanol/water/7N ammonium hydroxide (600:200:5:3). Then the fraction containing the glyco-dihydroxyconjugates was separated by rechromatography using acetonitrile/acetic acid (40:1) as mobile phase, and the fraction containing the tauro-dihydroxy conjugates was partially resolved by rechromatography with an eluent of acetonitrile/acetic acid/water/97 per cent aq. formic acid (100:2:2:1). The *p*-nitrobenzyl esters of free and glycine-conjugated bile acids have been analyzed by HPLC on a column of MicroPak-NH$_2$ using gradient elution [47]. UV detection at 254 nm was used. Taurine conjugated bile acids could be separated on a μBondpak C_{18} column using a mobile phase of methanol/0.01M potassium dihydrogen phosphate (75:25) and detected at 210 nm [47].

9

Prostaglandins

Several investigations of the application of HPLC to prostaglandin (PG) analysis have been reported. Once again the major problem is one of detector sensitivity, especially when assays of biological materials are required.

Because 9-keto prostaglandins are converted by base into the conjugated B prostaglandins ($\lambda_{max} = 278$ nm) a UV detection system can be useful in certain cases. PGB_2 has been determined by chromatography on a 1.15 m x 1.8 mm i.d. column of SIL-X with a mobile phase of chloroform/ethyl acetate/formic acid (84.7:15:0.3) followed by UV detection at 280 nm [471]. With an eluent flow rate of 17 ml h^{-1} PGB_2 eluted in 18 min and 2–5 ng could be detected at a sensitivity of 0.01 AUFS. The recovery of PGB_2 from the column was quantitative, indicating the possibility of preparative applications. PGA_2 eluted slightly later than PGB_2 but the two compounds were not completely resolved. Under the same conditions PGE_2 and $PGF_{2\alpha}$ were retained on the column for at least 1 h, but they could be rapidly eluted by methanol. Their elution was followed by bioassay of eluent fractions. The method was applied to the analysis of prostaglandins produced by rat renal papilla homogenates. The isolated prostaglandins were treated with base and analyzed as PGB. The results obtained for prostaglandin synthesis by the HPLC method compared favourably with those obtained by bioassay [471].

The separation of PGA_2 and PGB_2 has been achieved by ion exchange chromatography on triethylaminoethyl cellulose or a pellicular strong anion exchanger using aqueous tris(hydroxymethyl)methylamine acetate solvents [472]. This method was also used to monitor the base catalyzed conversion of PGE_2 to PGA_2 and PGB_2 with detection being performed at 254 nm.

Using a column of pellicular silica, Andersen and Leovey separated closely-related prostaglandins such as PGE_0 (dihydro PGE_1), PGE_1 and PGE_2; $PGF_{0\alpha}$, $PGF_{1\alpha}$ and $PGF_{2\alpha}$; 13,14-dihydro $PGF_{2\alpha}$, $PGF_{1\alpha}$ and 5,6-*trans*-$PGF_{2\alpha}$ [473]. Microgram amounts of the prostaglandins were resolved and detected using an RI monitor.

The separation of the epimers $PGF_{2\alpha}$ and $PGF_{2\beta}$ has been achieved by chromatography on silica coated with 5 per cent silver nitrate with a mobile phase of toluene/dioxan/acetic acid/(–)-2-methylbutanol (30:30:2:1) [422]. PGA_1 and PGB_1 were also well resolved in the same system.

An HPLC method has been developed which results in a single-step chromatographic purification of PGA_2, PGE_2 and $PGF_{2\alpha}$ from blood plasma in about 1 h [474]. The samples were recovered in a state of purification suitable for sample analysis by GLC/MS. Chromatography was performed either on a bonded Carbowax 400 phase or on μPorasil using a gradient elution system in both cases. With slight modification these methods were also found useful for the analysis of prostaglandins from seminal fluid, tissues and urine [474].

Several recent publications have used prostaglandin derivatization to enhance the UV absorption of the compounds and increase the sensitivity of the analytical method [475, 476]. Morozowich and Douglas prepared p-nitrophenacyl esters by reaction of the prostaglandins with p-nitrophenacyl bromide in acetonitrile, using diisopropylethylamine as a catalyst [475]. The reaction was complete after 15 min at room temperature, and aliquots of the reaction mixture could be directly injected onto the column. No degradation or epimerization of the E and A prostaglandins occurred during the derivatization. At 0.001 AUFS, 1 ng of the p-nitrophenacyl ester of PGE_2 could be detected (signal:noise = 2:1). Some impressive separations included the resolution of a mixture of the esters of ten F-series prostaglandins in 20 min using two 250 x 2.1 mm i.d. columns of Zorbax SIL with a dichloromethane/hexane/methanol (55:45:5) mobile phase as shown in Fig. 9.1. Since the esters have a UV absorbance maximum near 254 nm this wavelength was convenient for detection. The p-nitrophenacyl esters of the E-series prostaglandins required a less polar solvent system and a mixture of eight components was resolved in 36 min using dichloromethane/acetonitrile/ dimethylformamide (160:40:1) as eluent. A complete separation of the esters of the isomeric prostaglandins PGA_2 and PGB_2 was obtained on one 250 mm column using the mobile phase hexane/dichloromethane/dimethylformamide (120:80:1), although very little separation of the two was observed on silica gel TLC. Another difficult TLC separation is that of the 15-hydroxy isomers of the 15-methyl prostaglandins, but the p-nitrophenacyl esters of the 15-(R) and 15-(S) methylprostaglandins of both the E and F series were easily separated by HPLC. The C-15 epimers of 15-methyl PGE_2 have also been resolved without prior derivatization on a column of μPorasil using a mobile phase of methyl acetate/acetic acid (99.5:0.5) [477]. Refractive index detection was used in this case.

Similarly 15-epi-$PGF_{2\alpha}$ has been determined in $PGF_{2\alpha}$ samples by chromatography on silica using a mobile phase of ethyl acetate/acetic acid (80:20) [478]. However, for the quantitation of 15-epi-$PGF_{2\alpha}$ in bulk drug samples of Dinoprost ($PGF_{2\alpha}$), p-nitrophenacyl esters were prepared [479]. The separation was obtained on a column of LiChrosorb SI60 using an eluent of acetonitrile/ dichloromethane/water (50:50:1). Using UV detection at 254 nm a limit of detection of 0.5 per cent was obtained for the 15-epimer.

The reversed phase liquid chromatograpic separation of closely related prostaglandins as their p-bromophenacyl esters has been reported [476]. A reversed phase system was employed to enable the repeated examination of

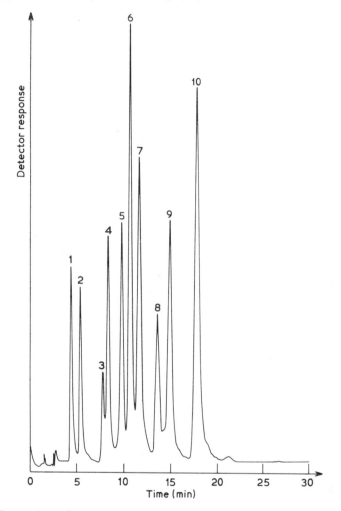

Fig. 9.1 Separation of *p*-nitrophenacyl esters of prostaglandins of the F-series. Column packing, Zorbax-Sil; column dimensions, 2 coupled 250 x 2.1 mm i.d.; eluent, dichloromethane/hexane/methanol (55:45:5); pressure, 3000 psi (20.4 MPa); flow rate, 0.3 ml min^{-1}; column temperature, ambient; detection, UV at 254 nm. Solutes 1, 13,14-dihydro-15-keto-PGF$_{2\alpha}$; 2, 15-keto-PGF$_{2\alpha}$; 3, 13,14-dihydro-PGF$_{2\alpha}$; 4, 15-epi-PGF$_{2\alpha}$; 5, 20-ethyl-PGF$_{2\alpha}$; 6, PGF$_{2\alpha}$; 7, 5-trans-PGF$_{2\alpha}$; 8, 11-epi-PGF$_{2\alpha}$; 9, 8-iso-PGF$_{2\alpha}$; 10, PGF$_{2\beta}$. Reproduced with permission from reference 475.

biological extracts without interference from endogenous material. The esters of PGF$_{2\alpha}$, PGE$_2$, PGD$_2$, PGA$_2$, PGB$_2$ and 15-methyl PGB$_2$ were separated in 45 min on a 250 x 4 mm i.d. column of μBondapak C$_{18}$ with a mobile phase of acetonitrile/water (50:50). Detection was performed at 254 nm and 50 ng injected on-column produced a 5 per cent response at 0.04 AUFS. The method was applied to the determination of micromolar levels of prostaglandins $F_{2\alpha}$, E_2 and D_2 formed from *in vitro* synthetase incubations.

$PGF_{2\alpha}$, PGE_2 and PGE_1 have each been separated from their major 15-keto metabolites by chromatography on μBondapak C_{18} after conversion to their *p*-bromophenacyl esters [480]. The detection and quantitation of prostaglandin in 1.0 ml of a 5 μM solution was possible, and the method was applied to the monitoring of the prostaglandins formed during *in vitro* incubations with prostaglandin 15-dehydrogenase from monkey lung.

10
Tricarboxylic acid(TCA)cycle and related organic acids

TCA cycle and related organic acids have been separated by adsorption chromatography [481], reversed phase chromatography [482] and ion exchange chromatography [483—486]. The problem of low sensitivity of detection has been overcome by the use of variable wavelength spectrophotometers monitoring at 210 nm for the -COOH chromophore [481, 487], or by the preparation of UV absorbing derivatives [482, 488]. A specific system devised for the analysis of carboxylic acids involved addition of the sodium salt of o-nitrophenol (a weaker acid than the TCA acids) to the eluent [125]. The carboxylic acids were thus neutralized and the disappearance of the o-nitrophenol anion (giving o-nitrophenol) was followed by monitoring at 432 nm, its absorbance maximum. A detector used by Katz *et al.* measured the fluorescence of Ce^{III} obtained by the reaction of oxidizable compounds with Ce^{IV} [484, 485]. This was used in series with a UV analyser, and low microgram quantities of hydroxy acids [484] and aromatic acids [485] were detected in body fluids. Carlsson and Samuelson used an automatic four-channel analyzer for the detection of hydroxy acids [486]. Four reactions were employed *viz*: (1) chromic acid oxidation, (2) a carbazole colour reaction, (3) periodate oxidation with formaldehyde detection and (4) periodate oxidation and determination of periodate consumed. A new efficient carboxylic acid analyzer employs anion exchange chromatographic separation of the acids followed by specific detection of the hydroxamic acids formed by the coupling reaction between the carboxylic acid and hydroxylamine [489, 490].

Several supports have been investigated for the separation of TCA cycle acids. Oxalic, lactic, α-ketoglutaric and trans-aconitic acids were separated in less than 20 min on a 1 m x 2.1 mm i.d. column of Corasil II using hexane/isopropanol (50:1) as eluent [481]. Succinic, glycolic, citric and fumaric acids were individually chromatographed on the same column using a mobile phase of hexane/isopropanol (100:3). Using a mobile phase consisting of the organic layer from the ultrasonic mixing of 0.1N sulphuric acid, chloroform and *t*-amyl alcohol (with the aqueous layer being used to hydrate the silica) partial separation of acetic, fumaric, pyruvic, glutaric, β-hydroxybutyric, lactic, succinic and α-ketoglutaric acids was achieved on a 600 x 3.2 mm i.d. column of silica (15–25 μm) [125]. The separation of α-ketoglutaric acid (which gave three peaks by decomposition), cis-aconitic, malic, citric and isocitric acids was obtained in 60 min with this system using a 300 x 3.2 mm i.d. column [242]. C_{18}/Corasil has been found to give poor resolution of TCA cycle acids and a dimethylsilanized silica (5 μm YANA-PAK SILS) required a high pressure drop to elute the samples in a reasonable time [487]. However, chromatography on LS-140 (a carboxymethyl polystyrene support) with a mobile phase of hexane/tetrahydrofuran/t-butanol (20:1:1) gave separations of acetic, formic, fumaric and pyruvic acids in 10 min, and acetic, formic, succinic, glycolic, l-malic, α-ketoglutaric, citric and cis-aconitic acids in 30 min using flow programming. UV detection at 220 nm was employed [487]. The separation of maleic, citraconic, fumaric, acetic and acrylic acids has also been achieved by ion exchange chromatography on a 230 x 9.0 mm i.d. column of Aminex 50W-X4 (30–35 μm) cation exchange resin with an eluent of 0.001N hydrochloric acid at 0.8 ml min^{-1} [483].

Several biologically significant dicarboxylic acids were converted to their phenacyl, p-nitrobenzyl, or p-bromobenzyl esters for UV detection and separated by reversed phase chromatography on a 250 x 4 mm i.d. column of C_9/Corasil II with a methanol/water mobile phase [482]. At 254 nm, 5 ng of the phenacyl or p-nitrobenzyl derivative of malonic acid and 15 ng of the adipic acid derivative were detected at a sensitivity of 0.02 AUFS.

Phenylpyruvic acid is excreted in large amounts in the urine of patients suffering from phenylketonuria. It has been determined by HPLC after conversion into 3-benzyl-2-hydroxybenzoquinoxaline by treatment with naphthalene-2,3-diamine, and extraction into carbon tetrachloride [488]. Chromatography was performed at 40° C on a 1 m column of Permaphase ETH using water/acetonitrile (72:28) as mobile phase at 0.5 ml min^{-1}. Addition of 2-mercaptoethanol suppressed formation of an interfering by-product and 2-chlorothioxanthone was a suitable internal standard. No interference was observed from 2-keto-acids such as pyruvic acid, 2-ketobutyric acid, 2-ketoglutaric acid or 4-hydroxyphenylpyruvic acid.

11

Carbohydrates

11.1 Introduction

Paper chromatography has been traditionally used for the analysis of sugars, sugar alcohols, and O-methyl sugars [491], and preparative scale separations have been achieved on columns of cellulose powder [492] and adsorbent charcoal [493, 494]. Efficient separations which depend on differences in the degree of polymerization (DP) of oligosaccharides have been achieved by exclusion chromatography [495—498] and ion exchange techniques have been widely used [499—512]. One frequently-used method is based on the ability of sugars to form complexes with borate ions which can then be exchanged on an anion exchange column [500—504]. Other workers have used various counter ions and elution with water or ethanol/water to accomplish partition separations of carbohydrates on ion exchange resins [505—512]. However, analysis by these methods, although fairly efficient, can be a lengthy procedure. A considerable improvement in analysis time has been achieved since the advent of pellicular ion exchangers and packing materials based on microparticulate silica.

Problems with detection systems for carbohydrates have limited the application of HPLC to carbohydrate analysis. Traditionally post-column colourimetric reaction detectors have been employed [507, 513—515] but more recently refractive index [516—518] and flame ionization [487] detectors have been used. Sensitive determination of carbohydrates has also been accomplished by the preparation of UV absorbing derivatives before chromatography [519, 520].

11.2 Separation of standard mixtures

11.2.1 Normal phase separations

(a) Sugars

The separation of several sugars on a column of LiChrosorb SI60 using ethyl formate/methanol/water mobile phases has recently been described [518]. As shown in Fig. 11.1, baseline separation of fructose, sorbitol, sucrose and lactose was obtained in less than 20 min using a mobile phase of ethyl

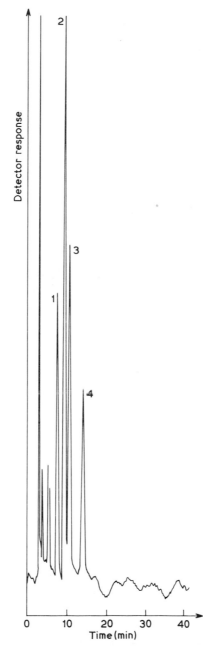

Fig. 11.1 Separation of sugars. Column packing, LiChrosorb SI 60 (5 μm); column dimensions, 150 x 4.6 mm i.d.; eluent, ethyl formate/methanol/water (6:2:1); pressure, 1030 psi (7.0 MPa); linear flow rate, 0.9 mm s^{-1}; column temperature, ambient; refractive index detection. Solutes 1, fructose; 2, sorbitol, 3, sucrose; 4, lactose; Reproduced with permission from reference 518.

formate/methanol/water (6:2:1). In this case refractive index detection was used. Improved detection levels have been obtained by the preparation of benzoate [519] or 4-nitrobenzoate [520] esters for use in conjunction with UV monitors. The separation of 16 perbenzoylated hydroxy compounds on Corasil II was achieved using a linear gradient of 0—99 per cent diethyl ether in hexane [519].

(b)   Glycols
Sequential degradation of a polysaccharide produces polyhydric alcohols, hydroxy ketones and aldehydes, hydroxy acids and sugar glycosides depending on the polyose structure. An HPLC method has been described for the separation and identification of some polyhydric alcohols and related compounds [521]. Separations were obtained on a column of Porasil A (75—125 μm) using a mobile phase of methyl ethyl ketone/water/acetone (85:10:5). The determination of ethylene glycol and diethylene glycol in polyethylene glycols, using propylene glycol as internal standard, was achieved by forming UV absorbing derivatives of the glycols with 3,5-dinitrobenzoyl chloride, and chromatographing the resulting derivatives on Corasil II, with heptane/ethyl acetate (75:25) as eluent [522]. Determinations were performed at the ng ml^{-1} level and higher molecular weight species were removed by back flushing.

11.2.2   Bonded phase separations
Several sugars have been resolved on Bondapak AX/Corasil using a mobile phase of acetone/ethanol/water (20:2:1) [487]. The separation of rhamnose, xylose, fructose, mannose and galactose was achieved in 30 min using this system. A flame ionization detector was used and detection of 5 μg of xylose was possible. The separation of galactose from glucose proved difficult on the column of Bondapak AX/Corasil but was achieved on a Vydac HC-SCX cation exchanger.

Mixtures of isomeric carbohydrates have been separated on a 1 m column of Pellosil-HC using dichloromethane as eluent [523]. The tribenzoates of methyl glycosides from D-glucose, D-galactose and D-allose were separated and the α-and β-anomers of methyl-D-glucopyranoside-2,3,6-tribenzoate, differing only in the configuration at one asymmetric centre, were resolved.

Several reports have been published on the use of amino-bonded phases for carbohydrate analysis [149, 170, 516, 517, 524]. Linden and Lawhead have described an automated carbohydrate analyzer which uses μBondapak Carbo-hydrate columns [517]. These columns were shown to be suitable for routine sugar analysis over prolonged periods. A column in daily use for four months showed only slight deterioration, requiring a mobile phase of gradually decreasing water content. The resolution of glucose, maltose, maltotriose, maltotetraose and maltopentaose from a starch hydrolysis sample using a mobile phase of acetonitrile/water (75:25) is shown in Fig. 11.2 [517]. The separation of oligo- and polysaccharides containing galactose and derivatives, frequently

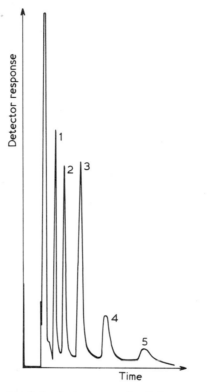

Fig. 11.2 Analysis of a starch hydrolysis sample. Column packing, μBondapak Carbohydrate (10 μm); column dimensions, 312 x 6.5 mm i.d.; eluent, acetonitrile/water (75:25); flow rate, 1.8 ml min^{-1}; column temperature, ambient; refractive index detection. Solutes 1, glucose; 2, maltose; 3, maltotriose; 4, malto tetraose; 5, maltopentaose. Reproduced with permission from reference 517.

found in sugar beet syrups, was also achieved in this system with a mobile phase consisting of acetonitrile/water (90:10). A similar method has been applied to the analysis of sugars such as glucose, maltose and maltotriose in hydrolyzed corn syrup [170]. In this case the mobile phase was acetonitrile/water (65:35). Oligomers of glucose from monomer up to DP 10, obtained from the hydrolysis of corn starch, could be separated in 30 min using flow programming [170].

Schwarzenbach used an aminopropyl bonded silica for the resolution of mixtures of neutral sugars, amino sugars and polyhydric alcohols [149]. The separation of ethylene glycol, glycerol, xylitol and sorbitol using a mobile phase of acetonitrile/water (80:20) is shown in Fig. 11.3. By reducing the water content of the acetonitrile/water mobile phase, the amino sugars could also be separated from the corresponding neutral sugars.

Another bonded phase which contains both amino and cyano groups has been developed and is sold as Partisil-10 PAC [525]. The separation of carbohydrates

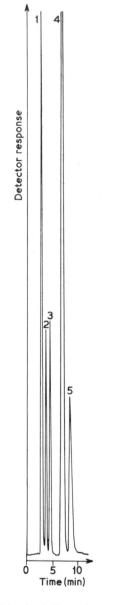

Fig. 11.3 Analysis of polyhydric alcohols. Column packing, aminopropylsilica; column dimensions, 250 x 3 mm i.d.; eluent, acetonitrile/water (80:20); flow rate, 0.5 ml min^{-1}; column temperature, ambient; refractive index detection. Solutes, 1, water; 2, ethylene glycol; 3, glycerol; 4, xylitol; 5, sorbitol. Reproduced with permission from reference 149.

on this phase has also been achieved with acetonitrile/water mobile phases, and the effect of the addition of acids or salts to the eluent has been examined [525]. Optimal k' values and peak shapes were obtained at pH 5 for most carbohydrates examined.

The analysis of aminoglycoside antibiotics such as the kanamycins has been considered in Section 6.2 and separations of various naturally occurring plant glycosides are discussed in Chapter 17.

11.3 Analysis of carbohydrates in body fluids

Glycoproteins are proteins with branching chains of carbohydrates that may be placed in three categories viz., amino sugars, neutral, and acidic sugars. The sugars are liberated from the glycoprotein by mineral acid hydrolysis. For the analysis of these liberated carbohydrates, it is important that the hydrolysis should be quantitative with minimal destruction of the carbohydrate. Neutralization and sample preparation should also involve no loss of carbohydrate and finally the detection system should have adequate sensitivity to quantitate the carbohydrates liberated. The determination of glycoprotein levels and of the protein/carbohydrate ratio in glycoproteins may be important in the diagnosis of cancer patients. A chromatographic system which was used for the analysis of glycoprotein hydrolysates consisted of a 1.5 m x 2.2 mm i.d. column of Bio Rad Aminex A-27 anion exchanger (12–15 μm) topped by a replaceable 30 mm column of anion exchanger and operated at a temperature of 55° C [63]. A 9-chamber gradient box was used to deliver concentration gradients of boric acid at 10 ml h^{-1}. The six neutral carbohydrates found in glycoproteins viz., mannose, fucose, arabinose, galactose, xylose and glucose were separated using a two-chamber gradient generator in 18 h with a borate gradient at pH 7. Using the nine-chamber box, 16 carbohydrates were detected with a cerate oxidimetric detector which monitored the fluorescence of Ce(III) produced by reaction of the solutes with Ce(IV). The detector was sensitive to 1 nmol of fucose.

Total protein was precipitated from the sera of 12 healthy women and after hydrolysis and neutralization, mannose, fucose and galactose were analyzed [63]. Separation was achieved in 4 h at 60° C on a 0.5 m column of Aminex A-6247 (2–10 μm). Fucose levels of 28–76 μg ml^{-1} of serum were found. In 22 out of 29 cases of women with disseminated breast cancer the protein-bound fucose levels were above the range of normal values [63].

Scott et al. have described equipment suitable for the analysis of carbohydrates and UV absorbing constituents of body fluids including urine, blood serum and cerebrospinal fluid [127]. The carbohydrates were detected by reaction of the column effluent with phenol/sulphuric acid and colourimetric monitoring at 480 and 490 nm. In the carbohydrate analyzer, up to 48 peaks were detected in a urine sample in 20 h using an anion exchange resin operated at 55° C with a borate buffer mobile phase. In urine from normal subjects, 30–40 carbohydrate peaks have been found in common, and peaks corresponding to, for example, sucrose, arabinose, galactose and glucose have been identified [526]. In samples from diabetic patients larger glucose peaks were observed [527]. For the detection of UV absorbing components a UV monitor was employed with an Aminex A-27 anion exchange resin using an acetate buffer at a temperature of $25°$ C for the first 12 h of the analysis and $60°$ C for the subsequent 28 h [127]. Usually 100—120 UV absorbing peaks were resolved in a 0.5—1.0 ml urine sample. Urinary profiles from normal individuals were similar with 60 major peaks common to all and 80 common to most. Significant differences were observed in pathological samples. The chromatograms of serum and cerebrospinal fluid samples showed fewer peaks than the urine chromatograms [127]. Scott *et al.* have also described the use of coupled anion exchange and cation exchange columns to speed up and increase the resolution of urine analysis [126]. Carbohydrate detection was possible by continuously mixing the eluent 1:1 with conc. sulphuric acid at $100°$ C which resulted in the production of chromophores with absorbance maxima at 290—310 nm. Dual wavelength monitoring was used at 254 and 306 nm. Organic acids and oxidizable compounds, in amounts down to a few nanograms, were detected by cerate oxidimetry [126].

An early separation of urine components was reported by Burtis [528] who separated over 90 UV absorbing components in 20 h on an Aminex BRX (12—15 μm) anion exchanger with a linear acetate gradient. By fractionating the urine into an acidic and a combined (basic + neutral) fraction it was shown that the 0—3 h region of the whole urine chromatogram correlated with the chromatogram of the (basic + neutral) fraction, while the components in the acidic fraction eluted in the 3—20 h region of the whole urine chromatogram.

Identification and quantitation of the saccharides in rat urine was obtained by reversed phase chromatography on a μBondapak Carbohydrate column with an acetonitrile/water mobile phase and refractive index detection [529]. Differences were detected between the urinary profiles of normal rats and those of rats with pathological conditions.

12

Biogenic amines, amino acids and proteins

12.1 Catecholamines and their metabolites

Catecholamines have been analyzed by HPLC using ion exchange techniques [530–534], adsorption chromatography [535–538], ion pair partition chromatography [538–540] and recently by the technique of soap chromatography [538]. The best efficiencies were obtained in the ion pair partition and soap chromatographic modes [538, 539].

Persson and Karger [539] separated a range of biogenic amines by ion pair partition chromatography on a column of silica (250 x 3 mm i.d.) coated *in situ* with $0.1M$ perchloric acid using a mobile phase of tributyl phosphate/hexane (65:35) or ethyl acetate/hexane/tributyl phosphate (72.5:17.5:10). The amines were detected with a UV monitor at 280 nm. Similar separations were achieved with eluents of butanol/dichloromethane [538, 539].

Very good efficiencies were obtained for the chromatography of catecholamines on ODS/TMS silica using eluents containing small amounts of a suitable detergent such as sodium 1-dodecane sulphonate or sodium lauryl sulphate [538]. As shown in Fig. 12.1. fast separations of mixtures of biogenic amines and their metabolites are possible with soap chromatography. The method was applied to the analysis of catecholamines and their metabolites in urine, in particular pathological urines which contained larger amounts of catecholamines than normal.

A sensitive method for the analysis of adrenaline and noradrenaline in plasma was achieved by converting the amines into their tritiated methoxy derivatives, i.e. metanephrine and normetanephrine, which could be detected by liquid scintillation counting [540]. The amines were separated from interfering constituents by chromatography on a column of LiChrosorb SI 100 (10 μm) using a perchloric acid stationary phase and a mobile phase of dichloromethane/ isobutanol (60:40).

Increased sensitivity of detection has also been achieved by the preparation and chromatography of fluorescent catecholamine derivatives [535, 536, 541]. The fluorescamine derivatives of dopamine, norepinephrine and their 3-*O*-methyl metabolites, 3-methoxytyramine and normetanephrine were eluted from a column of Hitachi 3011 gel with methanol/$0.5M$ Tris-hydrochloric acid buffer,

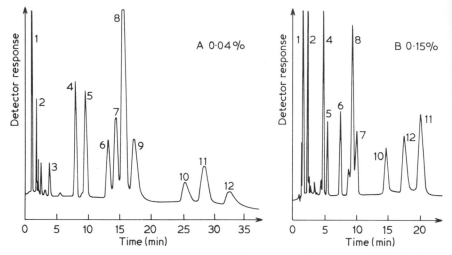

Fig. 12.1 Analysis of catecholamines by soap chromatography. Column packing, ODS/TMS silica; column dimensions, 125 x 5 mm i.d.; eluent, water/methanol/sodium 1-dodecanesulphonate (72.5:27.5:0.02, v/v/w) with added sulphuric acid: (A) 0.04 per cent (v/v); (B) 0.15 per cent (v/v); pressure, 1000 psi (6.8 MPa); flow rate, 1.3 ml min^{-1}; column temperature, ambient; detection, UV at 280 nm; sensitivity, 0.02 AUFS. Solutes, 1, homovanillic acid; 2, vanilmandelic acid; 3, 3,4-dihydroxyphenylglycol; 4, noradrenaline; 5, adrenaline; 6, normetadrenaline; 7, l-dopa; 8, dopamine; 9, metadrenaline; 10 tyramine; 11, α-methyldopa; 12, 3-methoxytyramine. Reproduced with permission from reference 538.

pH 8.0 (70:30) [541]. The eluted fluorophores were measured at the 100 pmole level. In another method the fluorescamine derivatives of dopamine and noradrenaline were chromatographed on a 250 x 2 mm i.d. column of MicroPak SI 10 using a mobile phase of benzene/dioxan/acetic acid (76:22:2) [536]. Alternatively, the dansyl derivatives of adrenaline, dopamine and noradrenaline were completely separated in less than 6 min on a 250 mm column of MicroPak SI 10 with a mobile phase of ethyl acetate/cyclohexane (60:40) [535]. Using fluorescence detection the amines could be determined down to 50 pg.

Several acidic metabolites of biogenic amines have been separated by both normal phase [539, 542] and reversed phase [543] ion pair partition chromatography. In the normal phase methods either tetrabutylammonium [539, 542] or ethyltributylammonium [542] was used as the counter ion in the aqueous stationary phase and the mobile phase consisted of butanol/dichloromethane (96:4) [539] or butanol/dichloromethane/hexane mixtures [539, 542]. Vanilmandelic acid, homovanillic acid and mandelic acid were separated by reversed phase ion pair chromatography on a column of silanized LiChrosorb SI 60 using a pentanol stationary phase and an aqueous mobile phase containing 0.1M tetrabutylammonium buffered at pH 7.4 [543]. Vanilmandelic acid and 3-hydroxy-4-methoxymandelic acid have also been separated by cation exchange chromatography [544], and recently a method for the determination of vanilmandelic acid and homovanillic acid in urine by HPLC has been described [545]. After extraction from urine the samples were chromatographed on a 500 x 3 mm i.d. column of Hitachi Gel No. 3010 (spherical 25 μm particles of polystyrene/divinylbenzene) at 30°C using a 0.05M tartrate buffer/methanol (80:20) mobile phase with a pH gradient, and detected at 280 nm. The minimum detectable concentration for both acids was 4 μg ml^{-1}.

12.2 Indoles

The analysis of indoles in urine has been achieved by ion exchange chromatography with fluorimetric detection [546–548]. Both acidic and basic compounds could be analyzed simultaneously by the use of coupled anion and cation exchange columns [546]. The indoles were eluted with an ammonium acetate/acetic acid buffer at 60°C. However, this system required 3 h per analysis and subsequently separate methods were developed for the analysis of indoleacetic acids [547] and serotonin and tryptamine [548]. The strong fluorescence of these compounds permitted their detection at the 100 ng level. Several indoles have recently been separated by chromatography on columns of Bondapak C_{18}/Corasil [547] or μBondapak C_{18} [549]. In the former system a mobile phase of methanol/0.2M phosphate buffer at pH 6.0 was used with gradient elution. In the latter case, after examination of the variation of k' with methanol content and pH, a mobile phase of water/methanol (90:10) at pH 4.6 was chosen for rapid analysis. The achievement of selectivity with fluorimetric detection permitted the analysis of the indoles after a single urine deproteinization step [549]. The detection limits were 5 ng for 5-hydroxytryptophan, 10 ng for 5-hydroxyindoleacetic acid and 15 ng for serotonin (signal/noise = 2:1). Indoleacetic acid and 5-hydroxyindoleacetic acid have also been determined in 1 ml urine samples by ion pair chromatography [540]. The extracts were analyzed on LiChrosorb SI 100 coated with 0.1M tetrabutyl-ammonium in Tris buffer (pH 8.3) using eluents of hexane/dichloromethane/ n-butanol. HPLC determinations of 5-hydroxytryptophan have been used as the basis for assays for tryptophan hydroxylase [550, 551]. These methods are discussed in Section 12.4 under amino acids.

12.3 Diamines and polyamines

Diamines and polyamines are considered to be important for the control of cellular growth, and enhanced urinary excretions of these amines have been observed in patients with different types of cancer [552]. Several analytical methods based on HPLC have been developed for their analysis, including ion exchange chromatography [530, 553–558], ligand exchange chromatography [559] and the normal phase [560] or reversed phase [561, 562] chromatography of derivatives.

In ligand exchange chromatography [563], a metal ion is bound to the support by electrostatic forces (e.g. to an ion exchange resin) or by coordination (e.g. to a chelating resin containing iminodiacetate groups). The ligands to be resolved (e.g. amines, alcohols, amino acids etc.) which are present in the eluent displace weakly-held solvent molecules (e.g. H_2O, NH_3) from the coordination sphere of the metal ion and, assuming this process to be reversible, are resolved according to the strength of the bond(s) which they form with the metal ion. Ligand exchange chromatography with RI detection provided a simple means of analyzing mixtures of diamines and polyamines with minimum interference from accompanying amino acids [559]. By varying the nature of the metal ion and ion exchange resin a great variety of elution orders could be realized. The best separations were obtained with a zinc-loaded sulphonated polystyrene resin (Aminex A-7) and with a copper-loaded acrylic type resin (BioRex 70) using aqueous ammonia eluents.

Fluorescamine derivatives of aliphatic diamines and polyamines have been separated on a Vydac reversed phase column at 60° C with a methanol/borate buffer (pH 8.0) gradient system [561]. Using fluorimetric detection the limit of detection was at the pmole level. As shown in Fig. 12.2, the dansyl derivatives of spermine, spermidine, cadaverine (pentane-1,5-diamine), putrescine (butane-1,4-diamine), propane-1,3-diamine and ethane-1,2-diamine were completely resolved in 6 min on a 250 x 2.2 mm i.d. column of MicroPak Al-5 using a mobile phase of chloroform/isopropanol (100:1) [560]. A similar method, using gradient elution of the dansylated amines from a column of MicroPak CN-10,

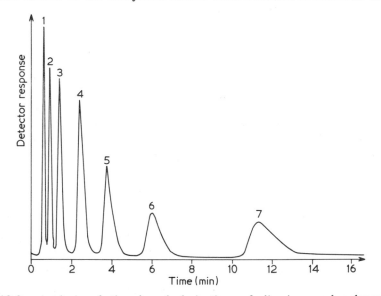

Fig. 12.2 Analysis of the dansyl derivatives of diamines and polyamines. Column packing, MicroPak Al-5; column dimensions, 250 x 2.2 mm i.d.; eluent, chloroform/isopropanol (100:1); flow rate, 1 ml min^{-1}; column temperature, ambient; detection, UV at 280 nm. Solutes: dansyl derivatives of 1, spermine; 2, spermidine; 3, cadaverine; 4, putrescine; 5, 1,3-diaminopropane; 6, 1,2-diamino-ethane; 7, ammonia. Reproduced with permission from reference 560.

was used to determine the levels of putrescine, spermidine and spermine in various tissues of rats and in mouse L 1210 leukaemic cells grown in culture [564]. The fluorimetric detection method was sufficiently sensitive to detect 40 pmole of putrescine and 20 pmole of spermidine and spermine. Putrescine, spermidine and spermine have also been determined as their tosyl derivatives on a 1 m x 2.1 mm i.d. column of Zipax Permaphase ETH [562]. The derivatives were chromatographed at 30° C using a water/acetonitrile (60:40) mobile phase and were detected at 254 nm. Nine replicate determinations on an identical mixture containing putrescine (6 μg), spermidine (9 μg) and spermine (30 μg) gave coefficients of variation of 2.75 per cent, 0.90 per cent and 2.57 per cent respectively.

The liquid chromatographic conditions required for the separation of approximately 50 aromatic monoamines and diamines have been established [565]. The separations were carried out on various silica gels using chloroform and/or cyclohexane mobile phases.

12.4   Amino acids

12.4.1   Ion exchange chromatography
A considerable amount of work has been published on the ion exchange chromatography of amino acids. Since the pioneering work of Spackman et al. on the separation of all 20 protein amino acids by automated ion exchange chromatography [61] the technique has been constantly improved and refined [6]. For example, the original analysis time of 21 h has been reduced to about 1 h [222].

High resolution ion exchange chromatography of amino acids has been achieved using a pH gradient derived from the accurate mixing of only two buffers of high and low pH with equimolar sodium concentration. Initially a citrate/phosphate buffer system was used [567] but subsequently a citrate/borate buffer system was found to give better control of the pH gradient in the basic region with a resulting improvement in the separation of the appropriate amino acids as shown in Fig. 12.3 [568]. Unger obtained an efficient separation of some 13 amino acids using a sulphonic acid cation exchange material and a stepped gradient system [569].

Rapid ion exchange chromatography of plasma amino acids has been achieved on polystyrene-based cation exchange resins [570]. Resins with 7.0–7.5 per cent divinyl benzene cross-linking were found to be the most satisfactory for the separation of the acidic and neutral amino acids present in physiological fluids, especially when lithium buffers were used. By using alkaline buffers containing borate ions with a pH gradient system all the amino acids could be eluted from one column.

12.4.2   Detection systems
UV detection is not universally applicable to amino acid determination and usually derivatives are formed either pre- or post-column to enhance their

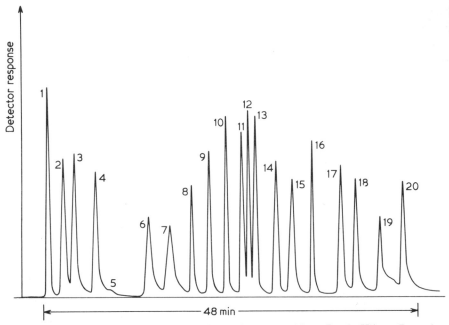

Fig. 12.3 Analysis of amino acids. Column packing, Rank Hilger 7 μm ion exchange resin; column dimensions, 350 x 2.6 mm i.d.; eluent, pH gradient from pH 2.2, citrate buffer $(0.2M \text{ Na}^+)$ to pH 11.5 borate buffer, $(0.2M \text{ Na}^+)$; pressure, 300–400 psi (2.0–2.7 MPa); flow rate, 0.16 ml min^{-1}; column temperature, 60° C; amino acid analyzer detector. Solutes: 1, aspartic acid; 2, threonine; 3, serine; 4, glutamic acid; 5, proline; 6, glycine; 7, alanine; 8, cysteine; 9, valine; 10, methionine; 11, isoleucine; 12, leucine; 13, norleucine; 14, tyrosine; 15, phenylalanine; 16, histidine; 17, tryptophan; 18, lysine; 19, ammonia; 20, arginine. Reproduced with permission from reference 568.

detection. A radioactivity detector for HPLC has been described [113] and the system has been used for amino acid analysis [571]. The resolution obtained was as good as that achieved with the ninhydrin detector [571]. Counting efficiencies for ^{3}H and ^{14}C were comparable with those obtained by static counting in vials.

(a)   Post-column reaction detectors
Amino acid analyzers traditionally incorporated a post-column reaction detector in which the amino acids were reacted with ninhydrin to form a coloured derivative [564]. More recently, fluorimetric detection systems utilizing a post-column reaction with fluorescamine have been developed [124, 572]. Such a system was used to monitor the separation of sixteen common amino acids in 1 h [124]. A modified chromatograph with an automatic gradient generating system was used with a column of Durrum DC-4A strong cation exchange resin. A linear response was obtained from 100 pmole to 5 nmole. A comparison of the HPLC method and the conventional method using an amino acid analyzer

with a ninhydrin reaction detector showed that shorter separation times and better sensitivity (500 pmole *vs* 100 nmole) were obtained by the HPLC method [124]. Since the fluorescamine is added to the column effluent, the derivatization does not affect the selectivity of the separation. However, chromatography of the fluorescamine derivatives of several amino acids on μBondapak C_{18} gave two incompletely resolved peaks, which were suggested to arise from the alcohol-lactone equilibrium shown in Fig. 12.4 [573]. On the basis of this evidence these derivatives would be unsuitable for pre-column preparation. Post-column reaction with *o*-phthalaldehyde and fluorimetric detection of the product has also been used for the determination of amino acids as discussed in Section 12.4.3 [551].

(b) Pre-column derivatization

Phenylthiohydantoin derivatives. Phenylthiohydantoin (PTH) derivatives of amino acids occupy a prime position in the sequence determination of polypeptides. Edman's degradation method for the analysis of peptides and proteins sequentially produces the amino acid residues as their PTH derivatives [574]. Several publications on the separation of PTH amino acids by HPLC have appeared [113—120]. All PTH-amino acids except PTH-arginine and PTH-histidine could be separated on a silica column [118]. Elution was performed with a concave solvent gradient from hexane/propanol/methanol (3980:11:9) to propanol/methanol (11:9) and the run was complete in 40 min. Eluted peaks of 2—5 nmole were easily detected by their UV absorption at 254 nm. Alternatively, as illustrated in Fig. 12.5, 11 PTH-amino acids could be separated in 25 min by liquid liquid partition chromatography on a column packed with MicroPak-CN [119].

Dansyl derivatives. Several dansyl derivatives of amino acids have been separated by liquid—liquid partition chromatography on silica dynamically coated with water-saturated dichloromethane [206]. Using a mobile phase of dichloromethane (water-saturated) containing 1 per cent acetic acid and 1 per cent chloroethanol the dansyl derivatives of isoleucine, valine, leucine, tyrosine, alanine, tryptophan, glycine, histidine and lysine were eluted in 22 min. More polar amino acids such as serine, glutamic acid etc. were not eluted within a

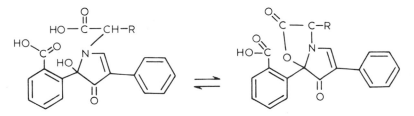

Fig. 12.4   Alcohol-lactone equilibrium of amino acid fluorescamine derivatives.

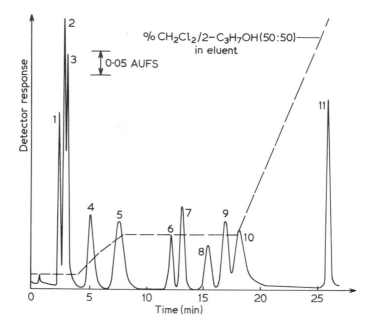

Fig. 12.5 Analysis of phenylthiohydantoin amino acids. Column packing, MicroPak CN (10 μm); column dimensions, 250 x 2.1 mm i.d.; eluent, gradient elution from hexane to dichloromethane/isopropanol (50:50); pressure, 670—1500 psi (4.6—10.2 MPa); flow rate, 0.86 ml min^{-1}; column temperature, ambient; detection, UV at 254 nm; sensitivity, 0.5 AUFS. Solutes: PTH derivatives of 1, proline; 2, leucine; 3, valine; 4, methionine; 5, alanine; 6, threonine; 7, glycine; 8, tyrosine; 9, lysine; 10, serine; 11, glutamine and asparagine. Reproduced with permission from reference 119.

reasonable time in this system. However, reasonable elution times were achieved for these amino acid derivatives by increasing the amount of chloroethanol in the mobile phase from 1 per cent to 10 per cent.

12.4.3 Applications: Amino acid assay methods

Taurine was rapidly separated from other strongly acidic amino acids on a 250 x 2.1 mm i.d. column of Aminex A-5 at room temperature with a mobile phase of 0.1M sodium perchlorate, adjusted to pH 2.2 with perchloric acid [551]. Detection was accomplished by post-column reaction with o-phthalaldehyde and fluorimetric detection of the product with excitation and emission wavelengths of 330 nm and 450 nm respectively. Using this method taurine was readily detected in a single rat pineal gland, and 2.0 nmole taurine could be determined, corresponding to 10 μmol g^{-1} wet weight. Other amino acids were strongly retained and tissue samples were repeatedly injected until column performance deteriorated. Regeneration was then effected by washing with 0.1M sodium hydroxide for 5 min and equilibrating the column at pH 2.2.

γ-Aminobutyric acid could be readily separated from interfering compounds on a cation exchange resin at pH 4.5 since its pKa (4.0) is higher than that of the α-amino acids (pKa 1.9–2.3) [551]. At this pH acidic and neutral amino acids eluted near the void volume and the aromatic and basic amino acids were more strongly retained. The effluent was mixed with o-phthalaldehyde and the γ-aminobutyric acid determined by fluorimetry as before. The peak height to sample size plot was linear up to 5 nmole and 50 pmole gave a signal:noise ratio of 10:1. Using this method, γ-aminobutyric acid was determined in rat substantia nigra, at a level corresponding to 0.11 μmole mg^{-1} protein [551]. An assay for glutamic acid decarboxylase was also developed by measuring both the endogenous γ-aminobutyric acid in the tissue extract, and the amount formed during the enzymic incubation, and subtracting one from the other. The activity of the enzyme sample was sufficient to increase the γ-aminobutyric acid content of the extract about six-fold.

5-Hydroxytryptophan does not occur in measurable amounts in the brains of normal animals but it accumulates when catabolism of the compound is blocked with a decarboxylase inhibitor. The rate of accumulation is an indirect measure of the turnover of the neurotransmitter 5-hydroxytryptamine. The same chromatographic conditions were used for the assay of 5-hydroxytryptophan as for γ-aminobutyric acid, except that a buffer at pH 5.0 was used [551]. Detection was by fluorimetry. The column effluent was mixed with concentrated perchloric acid pumped at 0.2 ml min^{-1} in order to take advantage of the acid-induced shift in emission of 5-hydroxytryptophan (from 360 nm at pH 5 to 550 nm in 3M acid). The technique was sufficiently sensitive to allow determination of 5-hydroxytryptophan accumulation in small nuclei (e.g. dorsal raphe) which have high tryptophan hydroxylase activity, and in large areas such as the hippocampus or corpus striatum where large amounts of tissue are available.

In another assay for tryptophan hydroxylase activity, tissue extracts were chromatographed on a 0.5 m x 2.1 mm column of Zipax SCX with a mobile phase of 0.1M sodium perchlorate at pH 2 [550]. The column effluent was mixed with concentrated perchloric acid and the fluorescence of the 5-hydroxytryptophan measured. A detection limit of 0.5 ng of 5-hydroxytryptophan in 50 μl extract was obtained.

12.5 Peptides and proteins

Until recently analysis of proteins involved enzymic degradation and determination of the protein hydrolysates on an amino acid analyzer. Several automated procedures have been developed [576–580]. Recently, progress has been made in the HPLC analysis of intact peptides and proteins and further developments can be expected in this difficult area.

Several polypeptide antibiotics [257, 263] and the chromapeptides constituting the actinomycin antibiotics [264] have been analyzed by HPLC. These separations have been discussed more fully in Section 6.2.

Some pharmaceutically important nonapeptides, oxytocin, lysin[8]-vasopressin and ornipressin, containing primary amino groups, have been separated by adsorption chromatography on LiChrosorb SI 60 using tetrahydrofuran/0.05M acetic acid (2:1) or by reversed phase chromatography on LiChrosorb RP-8 (10 μm) using water/acetonitrile buffered at pH 7 [581]. The samples were detected fluorimetrically after post-column derivatization with fluorescamine. The reversed phase system had the advantage of simpler sample preparation, better control of the derivatization reaction and optimization of solvent conditions. As a result detection limits of between 5—10 ng per injection could be obtained with a reproducibility of ±2 per cent.

Peptides from di- to decapeptide have been chromatographed on columns of Phenyl/Corasil, Poragel PN and Poragel PS under reversed phase conditions using acetonitrile/water mobile phases [582]. The presence of residual silanol groups on the Phenyl/Corasil material as well as the type of functional group on the Poragel (—OH and C=O on Poragel PN and aromatic-N on Poragel PS) was found to significantly influence the retention of the peptides. The separation of certain isomeric dipeptides has recently been accomplished using bonded optically-active tripeptide stationary phases [583]. Although the efficiencies of the columns were low, the separations obtained were superior to those which could be achieved on silica gel.

The controlled pore glasses developed by Haller [584] have the advantages of mechanical strength and chemical stability which make them potentially superior to dextran, agarose or polystyrene gels for exclusion chromatography. The main disadvantage encountered in their use is the strong adsorption of many proteins to the glass matrix [585]. This adsorption has been overcome by coating the glass with polyethylene glycol [585, 586] or by chemically bonding γ-glycidoxypropylsilyl groups to the silica surface [587]. The latter support was used to separate a mixture of human serum proteins in 25 min by elution with 0.05M sodium phosphate buffer at pH 7 [587]. Similar coating techniques have been used with macroporous silica. It was found that chemical modification of the surface of Silochrom silica by treatment with γ-aminopropyltriethoxysilane eliminated irreversible protein adsorption [588]. The protein elution volumes were shown to be dependent on the salt concentration and the pH of the eluent. A mixture of cytochrome C, lysozyme and α-chymotrypsin were chromato-graphed on a 400 x 1.6 mm i.d. column in 2.5 h [588].

The first separation of isoenzymes by HPLC was accomplished on a 500 x 4.6 mm column of Vydac pellicular anion exchanger [589]. Of the three isoenzymes of creatine kinase (EC 2.7.3.2), MM, MB and BB originating principally from skeletal muscle, cardiac muscle and brain respectively, two forms (MM and BB) were separated using a step gradient on the anion exchanger after preliminary exclusion chromatography on Sephadex G50 to remove small molecules from the tissue extracts. The third isoenzyme apparently denatured on the column.

Polymerization of triglycidylglycerol on the surface of controlled pore glass

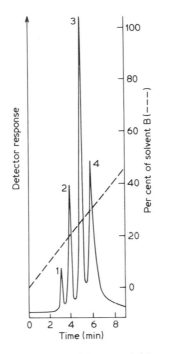

Fig. 12.6 Separation of a mixture of haemoglobins. Column packing, DEAE-Glycophase/CPG (25 nm pore diameter, 5–10 μm-particle size of CPG packing material); column dimensions, 250 x 4 mm i.d.; eluent, gradient elution from 0.0125M Tris, pH 8 to 0.0125M Tris/0.0125M Tris, 0.15M NaCl, pH 8 (Solvent B) (55:45); pressure, 2200 psi (15.0 MPa); flow rate, 2.5 ml min^{-1}; column temperature, 25° C; detection, UV at 410 nm. Solutes: 1, haemoglobin A$_2$; 2, haemoglobin S; 3, haemoglobin A$_1$; 4, haemoglobin F. Reproduced with permission from reference 591.

(CPG) supports with a glycerylpropylsilyl bonded phase produced a chromato-graphically stable glycerol polymer to which stationary phase groups were attached [590]. Diethylaminoethyl (DEAE), methyldiethylaminoethyl (QAE), carboxymethyl (CM) and sulphonylpropyl (SP) ion exchange supports of this type were suitable for the separation of proteins [590, 591]. The extremely useful separations which can be achieved on DEAE-glycophase/CPG is illustrated in Fig. 12.6 by the resolution of a haemoglobin mixture in 6 min on a 250 x 4 mm i.d. column at 25° [591]. A post-column enzyme reaction detector was developed whereby the eluent from the analytical column containing the enzyme was passed through a reaction column [591–593]. After introduction of an appropriate substrate for the enzyme, the incubation was performed in the reaction column and the product was measured spectrophotometrically. The resolution of the three creatine phosphokinase (CPK) isoenzymes and a mixture of lactic dehydrogenase isoenzymes on columns of DEAE-glycophase/CPG was monitored using this detector [591, 592].

13

Nucleotides, nucleosides and nucleic acid bases

13.1 Introduction

Since the pioneering work of Cohn on the application of ion exchange techniques to the separation of nucleic acid components [594, 595], nucleotides, nucleosides and nucleic acid bases have conventionally been separated by this method. Nucleotides, such as adenosine triphosphate (ATP, Fig. 13.1 I) are amphoteric and contain strongly acidic phosphate or pyrophosphate groups in addition to the nucleic acid base. Nucleotides are usually chromatographed by anion exchange chromatography [596—598] and a nucleotide analyzer based on anion exchange chromatography has long been known [599]. Nucleosides and nucleic acid bases such as adenosine (Fig. 13.1 II) and adenine (Fig. 13.1 III) respectively are usually analyzed by cation exchange chromatography in acid solution [596—598]. However, cation exchange chromatography has also been used for nucleotide separations [597], and Singhal and Cohn have separated nucleosides both by anion exchange chromatography in basic solution [600] and on a cation exchange resin at alkaline pH, using ion exclusion effects [601].

In the anion exchange separation of nucleotides, the monophosphates elute before the diphosphates which elute before the triphosphates. In the same system, nucleosides would elute prior to the monophosphates and the bases would be essentially unretained. In general, isocratic elution may be used for the separation of various monophosphates, but for a mixture containing mono-, di- and triphosphates, gradient elution (with a pH or ionic strength gradient) is usually required. The effect of eluent pH on the chromatography of the nucleotides is marked since the pH determines the degree of ionization of the phosphoric acid groups, as well as the protonation of the base moieties. The ionic strength of the eluent can also be adjusted to alter the retentions of the solutes, and improved efficiency may be obtained by operating at elevated temperatures.

Recently, nucleic acid bases and nucleosides [602, 603] and nucleotides [602, 604] have been separated by reversed phase chromatography. Interestingly, the elution order of adenosine mono-, di, and triphosphate (AMP, ADP and ATP) was the reverse of that obtained using anion exchange systems [602]. An alternative approach to the separation of nucleic acids and nucleotides has

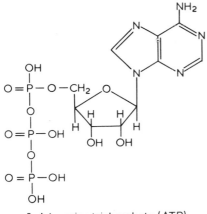

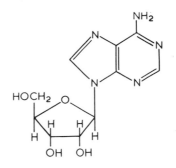

II. Adenosine

I. Adenosine triphosphate (ATP)

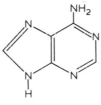

III. Adenine

Fig. 13.1   Structures of nucleic acid components.

been described [605–608]. In this method, a long chain quaternary amine is deposited on an inert support such as polychlorotrifluoroethylene beads [606, 608] or silanized silica [605] and the resulting phase used in conjunction with an aqueous buffer for nucleotide and oligonucleotide separations. A group separation of nucleotides, nucleosides, pyrimidine bases and purine bases has also been achieved on a polyacrylamide gel column [609].

Many of the published HPLC separations of nucleic acid components have been performed on rather inefficient ion exchange materials, such as those based on polystyrene/divinylbenzene resins. As shown, particularly by the work of P.R. Brown and co-workers, (discussed herein) much better efficiency can be achieved in these separations if modern microparticulate ion exchange or reversed phase materials are used. Since nucleotides and related molecules usually have molar absorbances of 10 000–20 000, detection of nanogram amounts of these compounds using UV photometers is possible [610]. HPLC techniques are thus well suited to the analysis of these compounds which are of great importance in biochemistry and of increasing importance in clinical chemistry.

13.2   Standard mixtures of nucleic acid components

Many separations of standard mixtures of nucleic acid components have been reported. Indeed, as this is the first step towards the identification of the components in more complex systems such as blood samples, most papers mentioned in the succeeding sections also contain standard chromatograms of nucleic acid components. Other examples include the pioneering separation of the mono-, di- and triphosphates of cytosine, uracil, adenine and guanine on a pellicular anion exchanger by Horvath and co-workers [11]. Later anion exchange separations of nucleotide mixtures were obtained on Aminex A-7 [611], Aminex A-27 [612], Aminex A-28 [613], Zipax SAX [614], Permaphase AAX [610] and Partisil-10 SAX [615]. Reports have also appeared on the analysis of nicotinamide [610, 615] and flavin coenzymes [615, 616], nucleotides in beef extracts [610] and beers [617], bases and nucleosides in beers [618], the N^1- oxides of AMP, ADP and ATP [619] and cyclic AMP in rat brain tissue [620] and rat liver homogenate [621]. Uracil derivatives have been analyzed at low levels (0.1 ng) using liquid–liquid partition chromatography [622] and cation exchange separations of standard mixtures of nucleic acid bases [614] and nucleosides [623] have been reported. Ribonucleotide reductase enzyme activity was monitored by HPLC on Aminex A-6 cation exchange resin. The method, which was highly reproducible [624], involved chromatography of nucleosides and gave similar results to a standard method.

13.3 Nucleic acid components in body fluids

13.3.1 Blood samples

The simultaneous analysis of cytosine, uracil, thymine, adenine, guanine, hypoxanthine and xanthine along with their nucleosides was achieved on Aminex A-28 strong anion exchange resin. An acetate buffer ionic strength gradient at pH 9.7 was used at a column temperature of 70° C [625]. This separation was applied to an *in vitro* study of purine metabolism in erythrocytes of a patient with Lesch Nyham Syndrome (LNS). When erythrocytes from a normal patient are incubated large amounts of guanosine triphosphate (GTP) are formed, whereas there is no significant change in the GTP concentration on incubation of erythrocytes from LNS patients. For normal patients the exogenous concentration of guanosine decreased and no other base or nucleoside was observed. With erythrocytes from LNS patients, as the guanosine concentration decreased the guanine concentration increased. Thus in agreement with enzyme research, guanosine was found to form guanine in LNS patients, but was not converted to GTP. (An increase in the urinary concentrations of the bases hypoxanthine, xanthine and uric acid in LNS patients has also been demonstrated by HPLC [626]). Nucleoside base profiles in blood etc. were also obtained [625].

Nucleotide profiles of cell extracts were obtained by HPLC on a pellicular ion exchange resin [627]. Peaks were identified by the use of internal standards, comparison with chromatograms of standard solutions, collection followed by chemical and spectrophotometric identification of fractions, and by the enzyme shift technique [627]. Chromatograms of extracts of erythrocytes, homogenized schistosomes or murine leukemia or sarcoma cells were obtained in 70 min. The coefficients of variation in the determination of 10 samples of a standard solution were 2.6 per cent and 1.5 per cent for AMP and GMP respectively. The total adenine nucleotide content in cell extracts by HPLC agreed with the enzymatic method in four different blood samples [627].

The nucleotide profiles of erythrocytes, leukocytes, lymphocytes, granulocytes and blood platelets were determined by chromatography on a 3 m x 1 mm column of Varian LFS pellicular anion exchanger with phosphate buffer gradients [628]. Reproducible profiles were obtained for the various elements using blood from normal patients. Erythrocytes were characterized by an ATP:ADP ratio of 7:1 and small amounts of guanine nucleotides, while platelets had an ATP:ADP ratio of 1.6:1 and a GTP:GDP ratio of 1:1. A plasma extract showed no appreciable concentration of nucleotides. The nucleotide profiles (e.g. the ATP:ADP ratio) of normal patients can then be compared with those of patients in diseased states or undergoing drug therapy [628].

Using a 1 m x 1.7 mm i.d. column of AS-Pellionex-SAX strong anion exchanger, Rao and co-workers found the ATP:ADP ratio in blood platelets to be 1.85:1 [629], a slightly higher value than that mentioned above [628]. Chromatography of platelet extracts on HS-Pellionex-SCX strong cation exchanger showed no appreciable quantities of nucleotide precursors or degradation products such as adenine, adenosine, guanosine, cytidine, uridine, inosine or hypoxanthine. In patients with Hermansky-Pudlak syndrome, the expected low concentration of platelet ADP was readily monitored by HPLC [629]. An advantage of the HPLC system is that AMP, ADP and ATP can be determined simultaneously, whereas with Firefly Luciferase which is commonly used for determining adenine nucleotide levels in platelets, only ADP and ATP are measured.

The production of guanine ribonucleotides in erythrocytes incubated with guanosine was monitored by HPLC on AS-Pellionex-SAX [630], and a column of Partisil-10 SAX was used in a study of the blood ATP levels of a patient with haemolytic anaemia [615]. Recently adenine and adenosine metabolism by intact human erythrocytes was studied by HPLC, using a column of AS-Pellionex-SAX for the separation of the nucleotides [631].

Nucleotides are usually extracted from the formed elements of blood using cold perchloric [629] or trichloroacetic [628, 632] acids, taking care not to hydrolyze the nucleotides. Scholar et al. have also pointed out [628] that the ATP:ADP ratio depends on the method used to separate the blood cells, and that a comparison of nucleotide ratios in normal and diseased states is valid only if the same separation method is used in both cases.

Recently, reversed phase chromatography on μBondapak C_{18} was used to separate more than 50 UV absorbing constituents in haemodialysate fluid from uraemic patients [633]. The method was more than ten times faster than ion exchange techniques and allowed detection of many materials at concentrations below 0.1 μg ml^{-1}. Compounds identified included caffeine, theophylline, xanthine and uric acid, etc. The determination of the level of the bases xanthine and hypoxanthine in plasma [634] and allopurinol and its major metabolite oxipurinol in serum at the $0.2 \mu g \, ml^{-1}$ level [635] have been reported. The analysis of theophylline in blood was discussed in Section 6.4.2.

13.3.2   Urine samples
The identification of over 100 individual compounds in urine samples has been referred to in Section 11.3. The method usually involves initial chromatography of a urine sample by anion exchange chromatography [611, 636] followed by re-chromatography of the fast eluting basic components on a cation exchange column [637]. Using this technique, xanthines and a range of bases (and acids) were identified in normal urine samples, and allopurinol and oxipurinol were two compounds additionally identified in pathologic urine specimens [637]. As mentioned previously (p. 156) HPLC was used to study the urinary excretion of UV absorbing compounds in patients with errors of purine and pyrimidine metabolism, such as the Lesch Nyham Syndrome [626].

13.4 Nucleotides and related compounds from ribonucleic acid (RNA) and deoxyribonucleic acid (DNA) samples
Samples of RNA (as little as $3-6 \mu g$) were hydrolyzed enzymatically to the nucleotide level with venom phosphodiesterase and subsequently to the nucleoside level with alkaline phosphatase. The fast analysis of the four nucleosides from RNA in nanomole and picomole amounts was achieved on a polystyrene-based cation exchange resin [638]. With a column temperature of $75°$ C and a pressure drop of 3000 psi (20 MPa) uridine (110 ng), guanosine (130 ng), adenosine (130 ng) and cytidine (120 ng) were separated in 4 min. Similar separations of the four DNA nucleosides were achieved [638]. The four deoxyribonucleotides obtained from a DNA sample by enzymic hydrolysis were analyzed in just over 1 h on Zipax SAX [639]. Using the long chain amine method mentioned in section 13.1, separations of oligonucleotides from RNA samples [606] and mono- and oligonucleotides from transfer RNA (tRNA) samples [608] were achieved. Individual tRNA species were also isolated on a preparative scale [608].

The analysis of methylated bases in tRNA is of interest since increased tRNA methylase activity has been reported in various cancer cells. A recent HPLC method was developed for the quantitative analysis of the four major tRNA bases along with 12 minor methylated bases [640]. The procedure involved gradient elution on Aminex A-27 anion exchange resin at pH 9.2. The minimum detectable level of the methylated bases was $0.1-0.2 \mu g \, ml^{-1}$ (signal:noise = 3:1) and the method was applied to the analysis of the methylated base content of calf and rat liver tRNA [640]. Aminex A-7 cation exchange resin has been used for the analysis of around 20 modified nucleosides from tRNA samples and cation exchange chromatography was also used for the quantitative analysis of 5-methylcytosine (as 5-methyl deoxycytidine) in DNA hydrolysates

[641]. The use of HPLC in the isolation of carcinogen-modified nucleic acid constituents (in this case, the reaction between deoxyguanosine and *N*-acetoxy-2-acetylaminofluorene) has been reported [642]. Also, in the *in vitro* alkylation of ATP with *N*-ethylnitrosourea the disappearance of the urea derivative and the appearance of alkylated ATP species were followed using anion exchange chromatography [619].

14

Porphyrins and bile pigments

Porphyrias are a group of metabolic diseases characterized by the abnormal production and excretion of porphyrins. Urinary porphyrins alone are not definitive with respect to the porphyrias but consideration of total urinary porphyrins and the pattern of faecal porphyrins allows easy differentation of these diseases.

Several recent publications [643–647] have described liquid chromatographic methods of obtaining the excretion patterns in the various types of porphyria which may facilitate clinical diagnosis.

Gradient elution of a methylated mixture of porphyrins, containing 4–8 carboxyl groups, isolated from the urine of a symptomatic porphyric patient was performed on a column of LiChrosorb SI 60 [644]. The solvent gradient was from hexane to ethyl acetate. Mesoporphyrin-IX dimethyl ester was used as internal standard and the porphyrins were detected at 400 nm. The technique was also shown to be applicable to the preparative scale separation of porphyrins. 10–15 mg quantities of the octa-, hepta-, hexa-, penta- and tetra-carboxylate fractions obtained from the faeces of poisoned rats were separated on a column of Porasil A-60 by gradient elution (hexane to ethyl acetate to methanol).

For the complete identification of a complex mixture of porphyrins, field desorption mass spectra were obtained from the eluted components [643, 644].

The patterns of urinary and faecal porphyrins in variegate-, symptomatic-, hereditary copro-, erythrohepatic- and congenital-porphyria were obtained by chromatography of the methyl esters on μPorasil using heptane/methyl acetate (60:40) as mobile phase, and flow programming [646]. The porphyrins were detected at 404 nm.

Carlson and Dolphin [647] initially fractionated the faecal extract by chromatography on silica under gravity to give a fraction containing porphyrins with 2–8 carboxyl groups and a 'sub-uro-porphyrin' fraction. Several systems were investigated for the resolution of the former material and it was found that Porasil T gave better resolution in the 4–8 carboxyl region. However, for the analysis of faecal samples in which the porphyrins containing 2–4 carboxyl groups dominated, Corasil II was preferred. The porphyrins were eluted with dichloromethane/light petroleum (30–60° C) (10:4) followed by dichloromethane/light petroleum/n-propanol (20:20:3) containing 10 μl per 100 ml triethylamine. Detection was accomplished at 254 nm. Initial results on the complex 'sub-uro-porphyrin fraction were promising. Monitoring of the eluent at 403.5 and 300 nm showed the presence of porphyrin material.

In the biosynthesis of uroporphyrinogen-III from porphobilinogen, one or more of four possible aminomethyl pyrromethanes, containing two carboxymethyl and carboxyethyl side chains, have been postulated as intermediates. During chemical synthesis of these intermediates, two stable derivatives of each pyrromethane were prepared and separation of both sets of isomers has been achieved by HPLC [648]. The four aminomethylpyrromethane lactam benzyl esters were completely resolved on a 600 x 4 mm column of 10 μm Porasil using dioxan/heptane (50:50) as eluting solvent with detection at 280 nm. The four aminomethyl pyrromethane α-unsubstituted lactams were also separated on the same column using ether/methanol (94:6) as mobile phase. The pure compounds could then be used for biosynthetic studies. Battersby investigated whether uroporphyrinogen produced by biosynthesis of aminomethylpyrromethane precursors was isomerically pure. The porphyrinogens were converted to the uroporphyrins for analysis but this proved unsuccessful. However, decarboxylation to the coproporphyrins was more useful. The coproporphyrins could be partially resolved as their tetraethyl esters into three fractions (I, II and III/IV) using reverse phase HPLC on two 300 x 4 mm columns of μBondapak C$_{18}$ (10 μm) with acetonitrile/water (70:30). The mixture of coproporphyrins III and IV was trans-esterified to the methyl esters which were then separated by recycle chromatography on μPorasil (10 μm) using ether/heptane (60:40) as mobile phase.

A wide range of synthetic bile pigments *viz.*, urobilins, violins and verdins of the IIIα, IXα and XIIIα isomer series have been separated as their dimethyl esters by adsorption chromatography on a 2 m x 3 mm column of Corasil II or a 300 x 4 mm column of μPorasil (10 μm) [649]. The urobilin separations were achieved with a basic solvent system consisting of benzene/ethanol/diethylamine (72500/2500/1) but the less basic verdins and violins required isooctane/methyl acetate combinations. The best separations were obtained on the microparticulate silica.

15

Vitamins

15.1 Fat-soluble vitamins

Fat-soluble vitamins have been separated by adsorption and reversed phase chromatography. Williams and co-workers have published a general paper on the quantitative analysis of vitamins A, D, E and K and their esters, using reversed phase chromatography on Permaphase ODS [650]. Ultraviolet detection is suitable for monitoring these compounds and a lower detection limit of 15 ng was obtained for vitamins A, D and K [650]

15.1.1 Vitamin A and the carotenoids

Compounds with vitamin A activity are formed in animals and humans by degradation of carotenoids (e.g. β-carotene, $C_{40}H_{56}$ and phytoene, $C_{40}H_{64}$) which are plant products. The *in vivo* cleavage of β-carotene (shown in Fig. 15.1) gives the aldehyde retinal, reduction of which yields vitamin A (retinol), the structure of which is also shown in Fig. 15.1.

Problems in the analysis of vitamin A centre on its ready oxidation, and its thermal and photochemical instability. Thus, GLC assays of silylated derivatives can cause decomposition or rearrangement; column chromatography of carotenoids is tedious and spectrofluorimetric methods are non-specific. The milder conditions of HPLC are more suited to the analysis of these compounds.

The presence of the four side chain double bonds in vitamin A means that chemical synthesis of the vitamin can give rise to various isomeric forms, each of differing biological activity. An HPLC method was developed for the separation of the various stereoisomers (as their acetate esters) in the presence of anhydro-vitamin A [651]. Liquid—liquid partition chromatography on 25 per cent BOP-coated LiChrosorb SI 60 (30 μm) was used with hexane as eluent. For the quantitative analysis of vitamin A or 13-*cis*-vitamin A in pharmaceutical formulations, adsorption chromatography on Corasil II was employed, along with a mobile phase of hexane/dioxan (99.9:0.1) [651]. The decomposition or isomerization of pure vitamin A caused by some batches of Corasil II was overcome by repeated injection of an antioxidant. Preliminary results indicated that improved chromatographic results could be obtained using microparticulate supports [651].

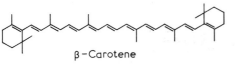 

β-Carotene   Vitamin A (all-*trans*)

Fig. 15.1

An HPLC method has been reported for the determination of blood levels of phytoene, which is under investigation as a sunscreen [652]. The pharmacokinetic characteristics of phytoene administered to dogs was followed by reversed phase chromatography of blood extracts on a 500 x 2 mm i.d. column of Permaphase ODS. Using a methanol/water (95:5) mobile phase, phytoene and β-carotene were separated in about 5 min. The recovery of phytoene from dog's blood was 86 per cent and the minimum level of detection was 50–100 ng ml^{-1} of blood [652]. Detection was accomplished at 280 nm where β-carotene has a relatively weak absorbance.

A study of the formation of vitamin A following incubation of β-carotene with tissue, was carried out by adsorption chromatography on Corasil II [653]. Radioactive vitamin A was isolated chromatographically after incubation of radioactive β-carotene. The HPLC analysis of vitamin A and β-carotene in foodstuffs has also been reported [654]. Alternatively, carotenoids have been separated by adsorption chromatography on magnesium oxide [655] and by exclusion chromatography [656].

15.1.2 Vitamins D₂ and D₃

Some impressive HPLC separations of vitamin D_2 (ergocalciferol) and vitamin D_3 (cholecalciferol) metabolites and isomers have been reported. As with vitamin A analysis, problems are encountered in the analysis of the D vitamins and related compounds due to their susceptibility to oxidation, and their thermal and photochemical instability. Adsorption and reversed phase chromatography have been applied to the analysis of these compounds.

Jones and DeLuca used two coupled 250 x 2.1 mm i.d. columns of Zorbax SIL to separate vitamin D_2 from its four known hydroxylated metabolites, and vitamin D_3 from its five known hydroxylated metabolites [657]. The latter separation is shown in Fig. 15.2. Using two different mobile phase conditions the corresponding metabolites of vitamins D_2 and D_3 were at least partially separated from each other. It was found that a 1α-hydroxy substituent had a much greater adsorptive interaction with the support than a side chain hydroxyl group. Thus, as shown in Fig. 15.2, the dihydroxy compound, 1α-hydroxy vitamin D_3 was more retained than the trihydroxy compound, 24,25-dihydroxy vitamin D_3. This factor facilitated the separation of 1α,25-dihydroxy vitamin D_3 from 25,26-dihydroxy vitamin D_3 in lipid extracts [657]. Radioactive vitamin D metabolites in rat blood and chick liver extracts were also chromatographed on this system after a purification step on a Sephadex LH 20 column [657]. The adsorption system did not separate vitamins D_2 and D_3 themselves [657], but a

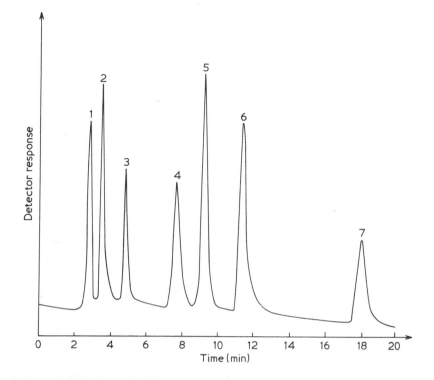

Fig. 15.2 Separation of vitamin D_3 from its five known hydroxylated metabolites. Column packing, Zorbax SIL (5 μm); column dimensions, two coupled 250 × 2.1 mm i.d. columns; eluent, hexane/isopropanol (90:10); pressure, 3000 psi (20.4 MPa); flow rate, 0.5 mol min^{-1}; detection, UV at 254 nm, sensitivity, 0–7 min, 0.02 AUFS, 7–20 min, 0.01 AUFS. Solutes: 1, solvent front; 2, vitamin D_3 (40 ng); 3, 25-hydroxy vitamin D_3 (30 ng); 4, 24,25-dihydroxy vitamin D_3 (25 ng); 5, 1α-hydroxy vitamin D_3 (40 ng); 6, 25,26-dihydroxy vitamin D_3 (40 ng); 7, 1,25-dihydroxy vitamin D_3 (25 ng). Reproduced with permission from reference 657.

partial separation was obtained on a Permaphase ODS column [650]. Recently, vitamins D_2 and D_3 have been completely resolved on a column of μBondapak C_{18} using an aqueous methanol eluent containing silver ions (341). Since vitamin D_2 has one double bond more than vitamin D_3, it forms a stronger silver complex and therefore elutes before vitamin D_3 in the reversed phase system. Separations of vitamin D_3 metabolites were also obtained using reversed phase chromatography on a 500 × 2.1 mm i.d. column of Permaphase ODS [658]. Using a methanol/water gradient system, the separation of 1α, 25- and 25,26-dihydroxy vitamin D_3 was achieved. About 10 ng of the 1α,25-dihydroxy compound was detected by monitoring at 254 nm. Resolution of the two compounds was better on the adsorption system described above [657]. The reversed phase method was also used to study the metabolism of tritiated 25-hydroxy vitamin D_3 to the 1α,25-hydroxy compound in chick kidney [658].

In a more recent publication, adsorption chromatography on a 250 x 2.1 mm i.d. column of Zorbax SIL was used to separate a wide range of vitamin D_3 metabolites and their trimethylsilyl derivatives [121]. Using dichloromethane/methanol (98:2) as eluent, separations of the vitamin D_3 metabolites similar to those discussed above [657] but with slightly better peak shapes, were obtained. In addition, epimers of the 24-hydroxyl group could be separated. Thus, for example, the tris-(trimethylsilyl) derivatives of 24R,25-dihydroxy- and 24S,25-dihydroxy vitamin D_3 were well separated using hexane/dichloromethane (98:2) as eluent. (The underivatized compounds were not separated.) This separation was used to assign the stereochemistry of the naturally occurring 24,25-dihydroxy vitamin D_3 metabolite [121]. The trimethylsilyl derivatives of 24R- and 24S-hydroxy vitamin D_3 were separated, and partial resolution of the free forms of $1\alpha,24R,25$-trihydroxy- and $1\alpha,24S,25$-trihydroxy vitamin D_3 was also achieved [121].

The photochemical synthesis of vitamin D_3 from 7-dehydrocholesterol results in a complex mixture of vitamin D_3 isomers. The HPLC analysis of such mixtures was carried out by adsorption chromatography on LiChrosorb Alox T alumina (5 μm) using chloroform as eluent [659], as well as on μPorasil silica using a mobile phase of chloroform/hexane/tetrahydrofuran (70:30:1) [660]. On the silica system, vitamin D_3 was separated from 7-dehydrocholesterol, lumisterol$_3$, tachysterol$_3$, previtamin D_3, iso-tachysterol$_3$ and dihydrotachysterol$_3$ [660]. The HPLC method was superior to biological and chemical methods for the correlation of the vitamin D_3 content of a resin with its antirachitic activity, since interfering compounds were separated in the chromatographic method [660].

Vitamin D_3 has been analyzed in the presence of vitamin A [661], vitamin A acetate [659] and in a mixture containing vitamin A acetate, vitamin A palmitate, vitamin E (α-tocopherol) and vitamin E acetate [660]. Vitamin D_2 was quantitatively analyzed in formulated products containing vitamin A acetate [662].

15.1.3 Vitamin E (tocopherols)

Vitamin E is a mixture of compounds known as the tocopherols, with α-tocopherol being the most common in pharmaceutical products. Chromatograms of a vitamin E mixture and vitamin E esters in tablet extracts have been obtained by reversed phase chromatography on Permaphase ODS [650] and a standard separation of vitamins A, D and E on μBondapak C_{18} has been reported [170]. The separation of α-, β-, γ- and δ-tocopherol was achieved by direct injection of plant oils onto a 1.5 m x 2.1 mm i.d. column of Corasil II [663]. The mobile phase was hexane/di-isopropyl ether (95:5) and fluorimetric detection was used. (The excitation and emission wavelengths were 295 nm and 340 nm respectively.) Quantitative recoveries were obtained for α- and β-tocopherols added to maize oil. α-Tocopherol has also been determined quantitatively in rat brain using a Corasil II column and fluorimetric detection [664].

15.1.4   Vitamin K

The reversed phase separation of a vitamin K analogue (menadione) from the other fat-soluble vitamins on Permaphase ODS has been reported [650], while vitamin K_3 was separated from the water-soluble vitamins B_2, B_{12} and C after chromatography on Corasil C_{18} [333]. The structurally related ubiquinones (quinones with an isoprenoid side chain) were resolved by reversed phase chromatography on μBondapak C_{18}/Porasil [665].

15.2   Water-soluble vitamins

Since the water-soluble vitamins possess basic or acidic sites, separations of these compounds are usually carried out by ion exchange chromatography. Williams and co-workers reported the chromatographic separation and quantitative analysis of most of the water-soluble vitamins and their phosphate esters including the analysis of the vitamins in multivitamin capsules and in foodstuffs [666]. Ultraviolet monitoring of the compounds at 254 nm or 280 nm usually gives satisfactory detection limits, except for vitamin B_5 (pantothenic acid) which has little UV absorbance.

15.2.1   The vitamin B group

The quantitative analysis of vitamins B_1 (thiamine), B_2 (riboflavin), B_6 (pyridoxine) and nicotinamide in multivitamin preparations was achieved on two coupled 500 x 2.1 mm i.d. columns of HS Pellionex SCX [667]. The influence of ionic strength and pH of the mobile phase on the chromatographic results was investigated. The quantitative analysis of vitamins B_1 and B_2 in foodstuffs using fluorimetric detection has also been reported [654]. Vitamin B_2 was separated from four of its photolytic degradation products by chromatography on a 1.22 m x 3 mm i.d. column of LiChrosorb SI 60 with an eluent consisting of chloroform/methanol/acetate buffer at pH 4 (60:20:4.5) [668]. In a conventional fluorimetric method for vitamin B_2 analysis, the photolytic degradation products interfered with the assay. The method was used to quantify vitamin B_2 in multivitamin tablets containing up to eleven vitamins as well as artificial colourings and flavours [668].

The individual members of the vitamin B_6 group (pyridoxine, pyridoxal and pyridoxamine) were separated by cation exchange chromatography on Aminex A-5 resin [669] and on Zipax SCX [666]. A polarographic detector was used to monitor the reversed phase separation of vitamins B_2, B_{12} C and K_3 [333].

Nicotinic acid (niacin) was quantitatively analyzed in vitamin tablet and foodstuff extracts [666] and a separation of nicotinic acid, isonicotinic acid and picolinic acid on Zipax SCX has been reported [670]. Nicotinic acid, folic acid and vitamin B_2 monophosphate were separated by anion exchange chromatography on Permaphase AAX, using a gradient elution system [666]. Ion exchange separations of radioactive nicotinic acid [671] and other B vitamins [569, 672] have been reported. Recently nicotinic acid in human serum (42 ng in 22 μl injected) was analyzed by reversed phase ion pair partition chromatography [673].

15.2.2 Vitamin C

In a review on automated vitamin analysis [674], the problems involved in the analysis of vitamin C (ascorbic acid) due to its ready oxidation were discussed.

The use of an amperometric electrochemical detector in the HPLC analysis of vitamin C in pharmaceuticals, body fluids and foodstuffs has been described by Pachla and Kissinger [675]. For the analysis of vitamin C in urine in 6 min, a 500 x 2.1 mm i.d. column of Zipax SAX was used with a 0.05M acetate buffer at pH 4.75 as eluent. Urine samples, diluted up to 1:100 with cold water, were injected directly into a pre-column which was used to protect the analytical column. The electrochemical detector was operated at an applied potential of 0.700 V vs a silver/silver chloride electrode. The system was capable of analyzing vitamin C in a wide range of multivitamin samples and foodstuffs. In foodstuffs where interferences from the matrix were common, the selectivity of the HPLC/electrochemical system was superior to the conventional iodimetric titration method. However, the method was not capable of analyzing vitamin C accurately at the normal serum levels of 5—15 μg ml^{-1} [675].

A reversed phase ion pair partition system has also been developed for the quantitative analysis of vitamin C [676]. A 300 x 4 mm i.d. column of μBondapak C_{18} was used, and the mobile phase was aqueous tridecylammonium formate (2 x 10^{-3}M) at pH 5.0/methanol (50:50). The resulting tridecyl-ammonium ascorbate ion pairs were eluted free from interferences in a wide range of multivitamin formulations and foodstuffs. Vitamin C levels of 5 μg ml^{-1} were readily measured using UV detection at 254 nm [676].

The ascorbic acid concentrations of aqueous humour samples of glaucoma eyes were determined by anion exchange chromatography [677]. Since the oxidized form of ascorbic acid (dehydroascorbic acid) does not absorb at 254 nm, peak identity was confirmed by following the disappearance and re-appearance of the tentatively-assigned ascorbic acid peak on oxidation and subsequent reduction with dithiothreitol [677].

Part IV
Environmental analysis by HPLC

16

Pesticides, carcinogens and industrial pollutants

16.1 Introduction

At present much emphasis is placed on monitoring the levels and effects of pesticides and carcinogens in our air, water and food supplies. Analytical surveys are undertaken to evaluate the persistence of pollutants and their metabolites in the environment with a view to defining acceptable daily intakes of pesticide residues and food additives. Two of the problems involved in this complex analytical undertaking centre firstly on the severe demands placed on the sensitivity and selectivity to be achieved by detection systems monitoring trace components, and secondly on the large number of samples which must be analyzed in pesticide screening projects.

Although the detectors currently available for HPLC do not usually match the sensitivity of GLC detectors (where the electron capture detector, for example, routinely monitors picograms of suitable compounds and the GLC/MS combination can detect about 100 pg of solute), the large sample volumes which can be injected onto HPLC columns means that the practical detection limits of the two methods are about equal [103, 678]. Examples of picogram amounts of solutes monitored in HPLC systems include 40 pg of aldrin (Fig. 16.1) using an electron capture detector [59], and 40 pg of aflatoxin G_2 [679] and 10 pg of aromatic hydrocarbons [680] using fluorimetric detection in both cases. Further, selective HPLC detectors such as spectrofluorimeters, electrochemical and electron capture detectors can significantly increase sensitivity or reduce sample clean up for suitable analytes. The advantages of wavelength variation in the UV monitoring of pesticides have also been described [44, 681]. Recently introduced continuous-flow radiometric detectors [682, 683] for HPLC also have an interesting future in investigations of the metabolism of radioactive pesticides. A highly efficient continuous-flow radioactivity detector for HPLC has been described by Reeve and Crozier [684]. In this system, column effluent was mixed with a liquid scintillator prior to scintillation counting and when used for the monitoring of radioactive gibberellin metabolites from plant extracts, counting efficiencies of 30 per cent and 80 per cent were obtained for ^{3}H and ^{14}C respectively [685]. Detectors for HPLC based on enzyme reactions have

Chlorinated insecticides

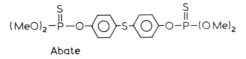

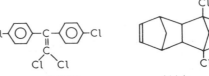

p,p′—DDT

p,p′—DDE
(metabolite of DDT)

Aldrin

Organophosphorus insecticides

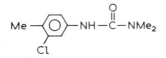

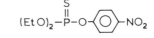

Abate

Parathion

Carbamates

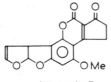

Carbaryl

Zectran

Benomyl

Substituted urea herbicide

Me —⟨O⟩—NH—C(=O)—NMe₂ with Cl

Chlortoluron

Carcinogens

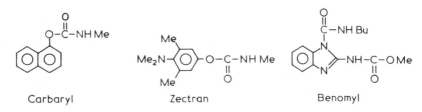

Aflatoxin B₁

Aflatoxin G₁

Benzo[a]pyrene

Fig. 16.1 Representative pesticides and carcinogens.

been described recently [592, 593, 686] and in one case [686] such a detector was used to monitor HPLC separations of organophosphorus and carbamate pesticides. Further advances in the HPLC/MS link-up can also be expected to improve detection limits and specificity in environmental analysis. The use of non-destructive detectors in HPLC also allows samples to be collected for structure elucidation or confirmation or, on the preparative scale, for biological testing or further chemical modification.

The solution to the problem of analyzing large numbers of samples in environmental screening projects (or in quality control), lies in the development of automated pesticide analyzers. Ott [687] has reviewed various types of automatic pesticide analyzers based on spectroscopic measurement, enzymatic reactions, TLC, GLC, atomic absorption and other techniques. An important advance in this area was made by Dolphin and co-workers who developed an automatic HPLC analyzer for monitoring organochlorine pesticides in raw milk [58]. (The system is discussed in more detail on p. 175.) The range of automatic injection devices for HPLC now available commercially will also reduce operator involvement in analyzing large numbers of samples.

As in other areas of analysis, HPLC is a useful complementary technique to GLC. An advantage of HPLC in environmental analysis is that molecules of varying polarity (e.g. pesticide + metabolite mixtures) can be analyzed in one chromatographic run. Since aqueous mobile phases can be used in reversed phase HPLC (including ion pair partition modes) sample preparation is often less extensive than in GLC. The number of clean-up steps can also be reduced by the use of a precolumn to protect the analytical column, or by a preliminary size separation of a crude extract on an exclusion column. The use of reversed phase HPLC in structure/activity correlations of pesticides is discussed in Appendix 7. Several reviews of HPLC in environmental analysis have appeared [687–690].

16.2 Pesticides

16.2.1 Insecticides

(a) Chlorinated insecticides and polychlorinated biphenyls

The analysis of chlorinated pesticides (Fig. 16.1) in residue samples is complicated by the fact that they usually occur along with polychlorinated biphenyls (PCBs). The latter compounds occur widely in the environment due to their use as plasticizers, dyestuff additives and hydraulic oils, and both types of compound are persistent in the environment. GLC analytical methods for the pesticides are frequently hampered by the co-elution of PCBs, although picogram amounts of the separated components have been detected using GLC with electron capture detection. Since both compound classes include non-polar aromatic molecules, adsorption chromatography has been the mode of choice for the HPLC separation of these compounds.

In a recent study of the chromatographic separation of 47 PCBs and 8 chlorinated pesticides [44] a 250 x 3 mm i.d. column of 5 μm LiChrosorb SI 60 silica was used with a dry hexane mobile phase. Using this simple system, hexachlorobenzene, aldrin, heptachlor, p,p'-DDE, o,p'-DDE, o,p'-DDT, p,p'-DDT and p,p'-DDD were separated from each other and from decachlorobiphenyl and biphenyl in 8 min. As shown in Fig. 16.2 a 10- to 80-fold increase in sensitivity was obtained for all pesticides except the DDE isomers (metabolites of DDT) by using the UV detector at 205 nm rather than 254 nm. Highly chlorinated PCBs such as Aroclor 1254, 1260 and 1268 (where 12 represents

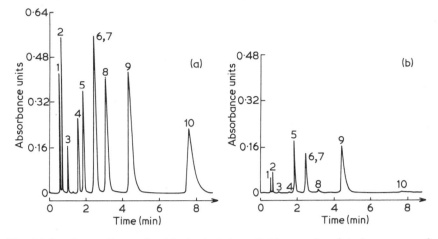

Fig. 16.2 Separation of chlorinated insecticides and related compounds. Column packing, LiChrosorb SI 60 (5 μm); column dimensions, 250 x 3 mm i.d.; eluent, dry hexane; flow rate, 4 ml min^{-1}; column temperature, 27° C; detection, UV at (a) 205 nm and (b) 254 nm. Solutes: 1, hexachlorobenzene; 2, decachlorobiphenyl; 3, aldrin; 4, heptachlor; 5, p,p'-DDE; 6, o,p'-DDE; 7, o,p'-DDT; 8, p,p'-DDT; 9, biphenyl; 10, p,p'-DDD. Reproduced with permission from reference 44.

biphenyl and the remaining two digits represent the per cent by weight of chlorine in the mixture) were shown to elute before the less highly chlorinated PCBs (Aroclors 1232 and 1248) and were quantitatively separated from longer-retained pesticides such as the DDT isomers, o,p'-DDE and p,p'-DDD. Pesticides with even longer retention times (lindane ≡ γ-benzene-hexachloride ≡ γ-BHC, endosulfan, endrin, dieldrin and methoxychlor) were well separated from all of the Aroclors. In a study of 47 individual PCBs and biphenyl itself, it was shown that increasing the number of chlorine atoms in the biphenyl nucleus led to a decrease in retention time, and the composition of Aroclor mixtures was interpreted tentatively on the basis of the retention data for individual components. The system was particularly useful for the separation of biphenyl from PCBs prior to quantitative analysis of PCB-containing mixtures by perchlorination procedures. Conversions of PCBs to decachlorobiphenyl and subsequent analysis by HPLC allowed detection of PCBs in concentrations down to 100 ppb (1 ng injected) *[44].

In an earlier paper [691] Aitzetmüller also separated the more highly chlorinated PCBs (Aroclor 1254 and 1260) from o,p'- and p,p'-DDT and partially from p,p'-DDE. The system used was again adsorption on silica

*In agreement with convention, concentrations in this section are quoted in terms of ppm where 1 ppm = 1 μg g^{-1} and 1 ppb = 1 ng g^{-1}. Where possible the lowest detectable concentration of pollutant in the matrix as well as the corresponding amount of pollutant injected onto the column are quoted.

(LiChrosorb SI 60, 10 μm) with dry petrol as eluent. Perisorb A, a pellicular, low capacity adsorbent, was used for the separation of PCBs as a group from o,p'-DDT and p,p'-DDT [678] and recently, commercially available poly-chlorinated naphthalene [692, 693] and PCB [694] mixtures have been invest-igated using HPLC on LiChrosorb (5 μm) columns with dry hexane as eluent.

An automated HPLC pesticide analyzer for the analysis of chlorinated insecticides and fat in milk was mentioned on p. 173. In this system, raw milk was injected onto a short silica precolumn where the fat was retained and the pesticides eluted with hexane [58]. The less polar, fast eluting pesticides such as p,p'-DDT, p,p'-DDE and α-BHC were not separated from each other by the precolumn and were stored at the top of a longer analytical column. Pesticides such as β-BHC, γ-BHC and dieldrin which eluted later were resolved from each other by the precolumn and were passed directly into an electron capture detector by means of a computer-controlled pneumatically-operated switching valve. Flow was then re-directed through the longer analytical column where the less polar pesticides were resolved before being monitored by the electron capture detector. The milk fat was removed from the precolumn by backflushing with hexane/isopropanol and the fat analyzed quantitatively using a refractive index detector. The total analysis time was 30 min and the HPLC/electron capture detector combination gave detection limits of 0.1 ppm pesticide in milk fat [58].

In contrast to the adsorption systems where the more highly chlorinated compounds elute before the less substituted analogues, a reversed phase system was described in which solute retention increased with increasing chlorine content [695]. Using a column of Vydac RP with methanol/water and ethanol/water mobile phases, a linear relationship was found for 17 pesticides between the logarithm of the adjusted retention relative to methoxychlor and the Beroza partition p-values. (The latter parameter is a measure of the percentage of solute distributed in the non-polar phase of a 1:1 pair of immiscible solvents. Values for a heptane/90 per cent aqueous ethanol mixture were taken to approximate most closely to a reversed phase chromatographic system of ODS/aqueous alcohol). The resulting linear relationship could be useful in choosing mobile phases for pesticide separations and in defining the polarity of unknown compounds in pesticide mixtures.

Silica gels bonded with dimethyldichlorosilane or ODS have been investigated for the clean-up step prior to the GLC analysis of DDT, DDE and lindane [696]. HPLC clean-up of environmental samples containing lindane prior to analysis by GLC has also been reported [697]. Fats and oils from water and fish samples containing lindane were rapidly eluted through a Corasil II column. Lindane was effectively separated from most of its GLC contaminants including α-lindane by this method. Recoveries of lindane in the presence of other pesticides such as DDT or DDD (which were removed by the Corasil II column) ranged from 92–98 per cent [697]. Adsorption chromatography on Porasil A columns has also been used for the separation of organochlorine pesticides from lipids prior to analysis of the pesticides by GLC with electron capture detection [698]. Separations of standard mixtures of aldrin, lindane, methoxychlor, DDT and related compounds were reported in early publications [197, 699–703].

(b) Organophosphorus insecticides

The organophosphorus pesticide Abate (Fig. 16.1) used to control mosquito larvae in natural waters, has been analyzed by chromatography on a 1 m x 2.1 mm i.d. column of Zipax coated with BOP [704]. Abate was extracted from pond water with heptane which was then reduced to a standard volume and injected onto the column, using heptane also as mobile phase. It was shown that an initial concentration of 100 ppb of Abate in pond water dropped over a period of about 10 h to a steady state concentration of 20–40 ppb. It was also shown that the majority of the Abate was adsorbed by organic matter in the pond and slowly released to the pond water over long periods of time. The level of Abate in extracts of dead larvae was measured as 1 ppm and similar results were obtained by placing freshly killed larvae into the injection system and effecting an instantaneous extraction by the application of high pressure [704].

Organophosphorus insecticides were separated on a 500 x 3 mm i.d. column of Permaphase ETH coupled to both a UV detector and an Autoanalyzer sensitive to cholinesterase-inhibiting compounds [686]. The advantage of this system is the specificity achieved – individual cholinesterase inhibitors are separated on the column and monitored in the Autoanalyzer which normally measures total cholinesterase activity. The coefficient of variation of the peak heights of various cholinesterase inhibitors was around 6 per cent, the rather high value being caused by the 30 min period required for the compounds to pass through the Autoanalyzer. The value is, however, acceptable in residue analysis. Organophosphates chromatographed and detected by this method included phosphamidon monocrotophos, CGA 18809 (an experimental insecticide) and dicrotophos. The latter two compounds (250 ng of each) were separated from each other and detected with this system. Two carbamate insecticides, dioxacarb and dimetilan, were also detected using this system. The limit of detection of CGA 18809 was 20 ng (signal:noise = 3) [686].

In a study of the analysis of organophosphorus pesticides, fenchlorphos, crufomate and fenthion were hydrolyzed to the corresponding phenols which were reacted with dansyl chloride to form fluorescent derivatives [705]. Separations of the three derivatives by adsorption chromatography on silica was complete in 8 min and 5–10 ng of the derivatives were detected. Adsorption chromatography on silica was used recently in a study of the chromatographic behaviour of 23 organophosphorus insecticides [706] and for the separation of the cis and trans isomers of the three organophosphorus pesticides dimethylvinphos, temivinphos and chlorphenvinphos [707]. Polarographic detectors have been used to monitor organophosphorus pesticides [51, 333, 708]. Using such a system, Huber and co-workers monitored residues of parathion (Fig. 16.1) in plant extracts at the 30 ppb level [708]. Recently HPLC separations of parathion and paraoxon [709, 710] methyl parathion [711] and pirimiphos methyl [712] have been reported.

(c) Carbamate insecticides

The carbamates, esters of carbamic acid ($NH_2 CO_2 H$), constitute an important family of insecticides of which carbaryl (Fig. 16.1) is the most common member. Carbaryl was determined in pesticide formulations on a Carbowax 400/Porasil column after prior extraction with methanol/chloroform and addition of pentachlorophenol internal standard [713]. The mobile phase was isoctane/chloroform (80:20). The 2-isomer of carbaryl (2-naphthyl-*N*-methylcarbamate) shows carcinogenic effects not exhibited by carbaryl itself, and hence this isomer must be absent from carbaryl formulations. A method for the analysis of the 2-isomer in the presence of carbaryl involved hydrolysis of the carbamates to 1- and 2-naphthol and analysis of the naphthols in the presence of unhydrolyzed carbamate on a Corasil II column with a hexane/chloroform (80:20) eluent [714].

Zectran (Fig. 16.1) a cholinesterase inhibitor and possible replacement for DDT in managing detrimental forest insects, hydrolyzes rapidly in alkali to 4-dimethylamino-3,5-xylenol. Photodecomposition of solid Zectran occurs at the 4-dimethylamino group, and an HPLC method for studying the compounds formed in aged Zectran solution employed reversed phase chromatography on ODS-SIL-X-II [715]. A liquid–liquid partition system was used for the analysis of methomyl insecticide, *S*-methyl-[(methylcarbamoyl)oxy]-thio-acetimidate [716].

Fluorigenic labelling has been employed by Frei, Lawrence and co-workers in the analysis of a wide range of carbamate insecticides [717, 718]. The carbamates were hydrolyzed in base to their corresponding amines which on reaction with dansyl chloride gave derivatives, 1–10 ng of which could be monitored using a fluorescence detector [717]. Chromatography of the derivatives was carried out on Corasil I [717, 718] and Zipax/BOP columns [717]. The analysis of water and soil extracts of the insecticide derivatives was possible without prior clean-up [717]. Apart from the advantage of selectivity shown by the fluorimetric detector which is useful in dealing with complex matrices, the detector is not flow sensitive [717]. Reviews are available on the various methods and reagents employed in the fluorigenic labelling of pesticides [48, 49].

Recently separations of around 30 carbamate pesticides by adsorption and reversed phase chromatography were reported. For many of the carbamates studied, improved detection limits (1–10 ng) were achieved by monitoring at 190–220 nm [681]. The detection of two carbamate insecticides, dioxacarb and dimetilan, using an Autoanalyzer sensitive to cholinesterase inhibitors has been mentioned (p. 176). This system had the advantage that no hydrolysis or derivatization steps were required. The carbamates chlorpropham and benomyl are dealt with in Sections 16.2.2, and 16.2.3 respectively.

(d) Pyrethrins

Pyrethrins are a group of naturally-occurring insecticides with low mammalian toxicity. An analytical scale separation of a pyrethrum extract on Permaphase ODS has been described [209] and the preparative separation of six pyrethrins has also been reported [181].

(e) Substituted urea larvicide

The substituted urea larvicide Thompson—Hayward 6040 [N-(4-chlorophenyl)-N'-(2,6-difluorobenzoyl)-urea] was analyzed in cows' milk following extraction and chromatography on Permaphase ODS with a methanol/water (50:50) eluent. Recovery of the larvicide from cows' milk was essentially quantitative and 10 ng amounts (a concentration of 100 ppb) were detectable [719]. The molecule was unsuitable for analysis by GLC due to either decomposition or adsorption on the GLC column. Residues of the compound in bovine manure were also analyzed by reversed phase chromatography on μBondapak C_{18} [720]. After prior extraction and clean-up by column chromatography, residues were quantitated at the 500 ppb level. (This corresponded to 200 ng injected and gave 25 per cent full-scale deflection on the recorder.)

(f) Insect juvenile hormones

Insect juvenile hormones are naturally-occurring compounds which retard the maturation of insects and which are therefore of interest as alternatives to chemical insecticides. Problems in the isolation of juvenile hormones centre on their low natural abundance and their instability. As has been pointed out in a recent review [721], an advantage of HPLC over GLC for the analysis of these compounds is that they can be recovered in good yield from HPLC columns in micro-preparative work. A study of the persistence of the synthetic juvenile hormone JH-25 [7-ethoxy-1-(p-ethylphenoxy)-3,7-dimethyl-2-octene] in flour samples was carried out by HPLC [722]. With UV monitoring at 226 nm (where the compound has an absorbance maximum) 10 ng were detected. Extracts of flour containing the active ingredient at the 0.5 ppm level were also quantitated. (Steroidal insect moulting hormones such as β-ecdysone have also been analyzed by HPLC [107]. The analysis of ecdysones by HPLC has been summarized in review articles [566, 723].)

16.2.2 Herbicides

(a) Substituted urea herbicides

The substituted urea herbicide chlortoluron (Fig. 16.1) was extracted from fortified soil samples with methanol and analyzed on a column of LiChrosorb SI 60 (10 μm) using hexane/isopropanol (85:15) as eluent. Following a silica gel column clean-up step the herbicide was analyzed quantitatively at the 100—200 ppb level in three soils [724]. Detection was carried out at 240 nm, the absorbance maximum of the herbicide. The method was not suitable for analyzing chlortoluron in the presence of monuron and diuron as these were not separated. A mixture of six substituted urea herbicides, diuron, neburon, linuron, monuron, phenobenzuron and isoproturon was completely resolved by chromatography on LiChrosorb SI 60 (5 μm) with a mobile phase of water-saturated dichloromethane [725]. The beneficial effect of the presence of water on the adsorption chromatography of the herbicides was discussed for both TLC and HPLC systems, the role of the water being to block the most active silanol sites thereby providing a less active but more homogeneous separating surface [725].

Recently three methods of herbicide monitoring (HPLC with UV detection at 254 nm; GLC with electron capture detection and GLC with electrical conductivity detection) were compared [726]. All three systems detected linuron at the 200 ppb level, although with the GLC/electrical conductivity system, larger amounts of samples had to be injected. With the HPLC/UV system, 3.4 ng of linuron gave 50 per cent full scale deflection at maximum sensitivity (0.005 AUFS). An early separation of linuron, diuron, monuron and fenuron was obtained by Kirkland using liquid—liquid partition chromatography [727] and reversed phase HPLC separations of a range of substituted urea compounds have been reported [728].

(b) Quaternary ammonium herbicides

Paraquat, a bipyridilium herbicide, was analyzed in clinical urine samples obtained from patients who had ingested the compound [34]. HPLC analysis of concentrations down to 200 ng ml^{-1} (10 ng paraquat injected) was possible by direct injection of untreated urine samples onto a column of alumina containing bonded aminopropyl groups. A methanol/aqueous acid eluent was used and paraquat was monitored by measuring the UV absorbance at 258 nm. Diquat, a related herbicide often found in combination with paraquat, was not detected at this wavelength. The system could also be used to separate the two herbicides, with diquat being detected at 310 nm. Good agreement between the HPLC method and a colourimetric method for the quantitation of paraquat was obtained [34]. Commercial paraquat formulations were also analyzed on Vydac cation exchange resin with 0.2M dimethylamine in methanol as eluent [729].

(c) Phenoxyacetic acid herbicides

The separation of five phenoxyacetic acid herbicides on a 1.5 m x 2 mm i.d. column of Perisorb A pellicular adsorbent with a mobile phase consisting of hexane/acetic acid (92.5:7.5) has been reported [678]. The substances are non-volatile and must be esterified before GLC analysis. With the HPLC system the detection limit of methoxychlorophenoxyacetic acid was 5 ng.

(d) Miscellaneous herbicides

Chlorpropham. A detailed study has been reported [730] of the analysis of the non-polar herbicide chlorpropham (isopropyl-3-chlorocarbanilate) and two polar hydroxy metabolites (2-hydroxy- and 4-hydroxy-chlorpropham) present as plant glycosides. Chlorpropham itself was extracted from alfalfa root and shoot tissue with chloroform, and the herbicide isolated by a combination of column chromatography and HPLC on Carbowax 400 with a hexane/chloroform (75:25) eluent. A fairly complex extraction procedure was used to isolate the polar chlorpropham metabolites from the alfalfa samples. The 2- and 4-hydroxy-chlorpropham conjugates could not be resolved by exclusion chromatography and thus enzymatic hydrolysis was used to convert the conjugates to the free hydroxy compounds which were subsequently acetylated. Although baseline resolution of the two compounds could be obtained using HPLC on Bondapak C_{18}/Porasil with an acetonitrile/water eluent, HPLC was used in the preparative mode to separate the acetylated derivatives from interfering plant material, and the acetates were then analyzed by GLC. Using radiolabelled chlorpropham it was possible, from radioactivity assays of the GLC peaks, to determine the ratio of 2-hydroxy:4-hydroxy metabolites in the original sample. The procedure was also extended to the analysis of rat and sheep excreta, from animals fed on herbicide-containing alfalfa shoot tissues [730].

Terbutryne. The herbicide terbutryne (2-ethylamino-6-methylthio-4-*t*-butyl-amino-1,3,5-triazine) was analyzed after extraction from fortified natural water samples by HPLC on Permaphase ETH with a water/methanol (80:20) eluent and by GLC using a conductivity detector [731]. The limit of detection of both systems was 1 ppb corresponding to 5 ng injected onto the HPLC column.

16.2.3 Fungicides

(a) Benomyl

Several papers have appeared on the analysis of the carbamate fungicide benomyl (Fig. 16.1). Kirkland *et al.* [732] used chromatography on Zipax SCX to analyze benomyl residues in soils and plant tissues after a preliminary liquid–liquid partitioning clean-up step. Extraction of benomyl from soils with acidic methanol converted benomyl to methyl-2-benzimidazolecarbamate (MBC), the principal degradation product of benomyl, and the MBC was analyzed quantitatively. A second degradation product, 2-aminobenzimidazole (2-AB) was eluted on the same chromatogram and likewise quantified. Recoveries of benomyl, MBC, and 2-AB from various soils averaged 92, 88 and 71 per cent respectively. Each of the compounds was monitored at the 50 ppb level. Ethyl acetate extraction was used for plant tissue analysis and a wide range of crops was satisfactorily analyzed. Fig. 16.3 shows the HPLC analysis of benomyl in cucumber extracts [732]. Kirkland also analyzed benomyl as MBC and two hydroxylated derivatives in cows' milk, faeces and tissues, again using Zipax SCX [733]. Fluorimetric detection was used to determine benomyl in plant tissues [734]. This detection system had the advantage that the clean-up step prior to the analysis was eliminated. The method was sensitive to benomyl at the 20 ppb level (20 ng of 2-AB injected).

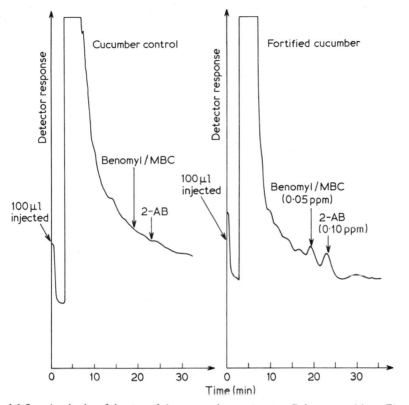

Fig. 16.3 Analysis of benomyl in cucumber extracts. Column packing, Zipax SCX pellicular strong cation exchanger; column dimensions, 1 m x 2.1 mm i.d.; eluent, 0.025N tetramethylammonium nitrate/0.025N nitric acid; pressure, 300 psi (2.0 MPa); flow rate, 0.5 ml min^{-1}; column temperature, 60° C; detection, UV at 254 nm; sensitivity, 0.02 AUFS. Solutes: benomyl (measured as MBC) and 2-AB at the concentrations shown. Reproduced with permission from reference 732.

(b)   Vitavax

Vitavax (5,6-dihydro-2-methyl-1,4-oxathiin-3-carboxanilide) is an anilide fungicide which produces two oxygenated derivatives on photolysis. Analysis of all three compounds was carried out on Bondapak C_{18}/Corasil with water/acetonitrile (80:20) as mobile phase [735]. After dichloromethane extraction of a fortified sample of lake water, 20 ppb of Vitavax (280 ng injected) and 5 ppb of its two photoproducts were determined by HPLC using UV detection at 254 nm. Cerium oxidation followed by fluorimetry was investigated as a detection system but responded only to Vitavax.

(c)   Dithianone

The fungicide dithianone was quantitatively analyzed after extraction from apples containing the compound at levels of 0.1 ppm and 20 ppb [678].

Injection of a standard solution containing 5 ng of dithianone gave a measurable peak (signal:noise = 2:1) when monitored at 254 nm.

16.2.4 Rodenticide

The anti-coagulant rodenticide warfarin [3-(α-acetonylbenzyl)-4-hydroxycoumarin] was determined in animal tissues, stomach contents, body fluids and feedstuffs [736] using a Corasil II column and an isooctane/isopropyl alcohol (98:2) eluent. The lower detection limit was 25 ppb, corresponding to about 25 ng of warfarin injected. Monitoring was at 270 nm. Two HPLC studies of the analysis of warfarin and its metabolites in blood using reversed phase chromatography have also been published recently [737, 738].

16.3 Carcinogens

16.3.1 Aflatoxins

Aflatoxins, metabolites of *Aspergillus flavus* are among the most potent known inducers of tumours. Aflatoxins from grain samples infested with *Aspergillus flavus* were extracted and chromatographed on a 1 m x 1.8 mm column of SIL-X silica with isopropyl ether/tetrahydrofuran (88:12) or diethyl ether/cyclohexane (75:25) eluents [739]. Aflatoxins B_1 and G_1 (Fig. 16.1) (1 μg of each injected) were resolved and monitored by UV detection at 254 nm. Peak identities were confirmed by off-line mass spectral analysis of the collected fractions [739]. In a study of the separation of aflatoxins B_1, B_2, G_1 and G_2, Seitz [740] found that Permaphase ODS, Phenyl/Corasil and Corasil II gave unsatisfactory resolution of the four compounds. However, the compounds were completely resolved on a 300 x 6.35 mm o.d. μPorasil (10 μm) column. Using 50 per cent water-saturated chloroform/dichloromethane (75:25) containing 0.5 per cent methanol, the elution order was B_1, B_2, G_1 and G_2 whereas with 50 per cent water-saturated dichloromethane containing 0.6 per cent methanol the order of elution of B_2 and G_1 was reversed. A UV spectrometer set at 350 nm was used to detect 25 ng of B_1 and G_1 and 7.5 ng of B_2 and G_2. This wavelength is more selective for aflatoxins and reduces interferences on the chromatogram from, for example, aromatics which could mask much of the chromatogram if detection at 254 nm were employed. A commercial fluorimetric detector exhibited much greater sensitivity for aflatoxins G_1 and G_2 (emission around 450 nm) than for B_1 and B_2 (emission around 400 nm), and the use of this detector was therefore limited for aflatoxin detection. A 50 μl injection of yellow corn extract containing 30 ppb of aflatoxin B_1 as determined by TLC, gave a value of 36 ppb when chromatographed on the μPorasil column with dichloromethane/methanol (99.2:0.8). The limit of detection was around 10 ppb [740].

HPLC was used to study *in vivo* and *in vitro* some oxidative metabolites of aflatoxin B_1 which is known to be oxidized by animal tissues to aflatoxins Q_1, M_1, B_{2a} and aflatoxicol H_1. All five compounds were separated in 20 min on a

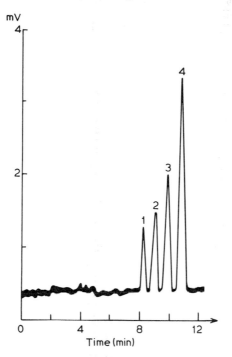

Fig. 16.4 Separation of aflatoxin standards. Column packing, μPorasil (10 μm); column dimensions, 300 x 4 mm i.d.; eluent, water-saturated dichloromethane/methanol (99.4:0.6); flow rate, 1.05 ml min^{-1}; column temperature, 20° C; detection system, fluorimeter with cell filled with purified silica (100–200 μm). Solutes: 1, aflatoxin B_1; 2, aflatoxin G_1; 3, aflatoxin B_2; 4, aflatoxin G_2 (2.4 ng of each). Reproduced with permission from reference 679.

240 x 2.1 mm i.d. column of Zorbax SIL with a mobile phase consisting of dichloromethane/chloroform (60:40), 50 per cent saturated with water and containing 0.9 per cent methanol [741]. Detection was by UV at 345 nm. Baseline resolution of all five compounds was achieved, in contrast to TLC systems in which B_{2a} and M_1 were only partially resolved. Around 5 ng of each compound could be detected at 345 nm by the HPLC method. With a fixed-wavelength broad-band fluorimetric detector the response of H_1 was enhanced but the other four compounds were barely detectable. Using the HPLC system, it was shown that M_1 was the principal metabolite excreted in the urine of rhesus monkeys orally administered aflatoxin B_1 [741].

Aflatoxins B_1, B_2, G_1 and G_2 were all resolved in 10 min on a 250 x 3 mm column of Partisil-5 silica (6 μm) with water-saturated dichloromethane/methanol (99.7:0.3) [742]. A variable wavelength detector set at 362 nm was used to monitor the samples and 10 ng of each aflatoxin was detected. In an independent study of the adsorption chromatography of the aflatoxins, 1–2 ng of each of the four aflatoxins was detected using UV monitoring at

360 nm [743]. Methods have also been reported for the HPLC analysis of aflatoxins in cotton seed products [744] and in wines [745]. The detection limits of the aflatoxins, B_1, B_2, G_1 and G_2 using fluorimetric monitoring were recently improved by packing the detector cell with a suitable adsorbent (100–200 μm silica) [679]. After adsorption chromatography, 120 pg of B_1, 80 pg of B_2, 120 pg of G_1 and 40 pg of G_2 could be detected with this system and for aflatoxin G_2 the plot of peak height *vs* amount injected was linear in the range 40 pg–40 ng. Fig. 16.4 shows the separation of 0.4 ng of each of the four aflatoxins with fluorimetric detection using a detector cell packed with silica [679].

16.3.2 Patulin
Patulin is a mycotoxin produced by moulds, particularly apple-rotting fungus and is a known mutagen and tumour promoter. An HPLC method for the analysis of patulin in foodstuffs was developed [746]. The technique involved the chromatography of extracts of patulin-containing apple juice on Zorbax SIL columns with an isooctane/dichloromethane/methanol (84:15:1) mobile phase. In a survey of 13 commercial samples of apple juice, patulin was found in 8 samples at levels of 40–300 ppb.

16.3.3 Phorbol ester
The diester of the terpene alcohol phorbol, phorbol myristate acetate, is the major active ingredient of croton oil and is a powerful irritant and tumour promoter. HPLC was used to study the thermal and photochemical stability of the compound [747].

16.3.4 Nitrosamines
The carcinogen *N*-nitrosonornicotine is found unburned in commercial U.S. tobaccos at levels of 0.3–90 ppm. The compound was conveniently analyzed on a 300 x 6 mm column of μPorasil with a chloroform/cyclohexane/methanol (68.6:30:1.4) eluent [748]. Partial separation of the *syn* and *anti* conformers of the carcinogen was also achieved on a μBondapak C_{18} column with an acetonitrile/water (50:50) mobile phase. With UV monitoring at 254 nm, 10 ng of *N*-nitrosonornicotine was detected. It was suggested that the method could be applied to the analysis of other nitrosamines in environmental samples [748]. The *syn* and *anti* conformers of some *N*-nitrosoamino acids and other *N*-nitroso compounds have also been separated by HPLC [749]. An HPLC method for the analysis of the carcinogen *N*-nitrosoproline involved denitrosation followed by fluorescent derivatization of the proline [750]. The sensitivity of the HPLC method (0.5 ng of the fluorophor detected) was around ten times higher than TLC and GLC/MS methods. (In practice, e.g. in the analysis of food samples by the HPLC method, any proline present would have to be removed prior to

denitrosation.) A nitrosamide-specific detector for HPLC has also been developed [751]. The use of HPLC to monitor nitrosamines has been discussed in recent conference proceedings [752].

16.3.5 Fluorenylhydroxamic acid esters

The separation and quantitative analysis of carcinogenic O-acetates of fluorenylhydroxamic acids was carried out using a column of Corasil II and a hexane/ethyl acetate (50:50) mobile phase [753]. The compounds were detected at 280 nm, and 100 ng of the acetate esters could be readily detected. The method was used to identify the previously unknown decomposition products of N-acetoxy-N-2-fluorenylacetamide in aqueous solution.

16.3.6 1,2-dimethylhydrazine

The separation of 1,2-dimethylhydrazine, a potent colon carcinogen in rats, from azomethane, azoxymethane and methylazoxymethanol, formaldehyde and methanol (probable metabolites of 1,2-dimethylhydrazine), was carried out on C_{18}/Corasil, μBondapak C_{18} and Aminex A-27, a strong anion exchange polystyrene-based resin [754]. Use of a μBondapak C_{18} column with 1 per cent aqueous ethanol eluent gave good resolution of all the compounds except 1,2-dimethylhydrazine itself which was excessively retained and badly tailed. This compound could be eluted in the protonated form, by changing the eluent to 0.05M acetic acid. Alternatively, the 1,2-dimethylhydrazine could be eluted rapidly through the Aminex A-27 column at pH 5.6, with reasonable resolution of the other metabolites (except methanol and formaldehyde) being achieved. Azo and azoxy compounds were detected at 205 nm and radioactive 1,2-dimethylhydrazine and formaldehyde were detected by liquid scintillometry of collected fractions. Radioactive metabolites in rat urine were separated on the Aminex A-27 column and analyzed by liquid scintillometry.

16.3.7 Benzo[a]pyrene

Benzo[a]pyrene (Fig. 16.1) is a powerful hydrocarbon carcinogen which is emitted into the atmosphere as a consequence of the combustion of fossil fuels. The compound is metabolized to oxygenated derivatives by the microsomal mixed-function oxygenases and studies of the analysis of these metabolites by HPLC have been reported. Using a 1 m x 2.1 mm Permaphase ODS column operated at 50° C and with a methanol/water gradient, benzo[a]pyrene was separated from 8 metabolites $viz.$, three dihydrodiols (9,10: 7,8 and 4,5), three quinones (benzo[a]pyrene-1,6-; -3,6-: and -6,12-dione) and two phenols (3- and 9-hydroxybenzo[a]pyrene) [755]. Using HPLC it was shown that 7,8 benzoflavone (an inhibitor of aryl hydrocarbon hydroxylase) inhibited the formation of all benzo[a]pyrene metabolites, while 1,2-epoxy-3,3,3-trichloropropane (an inhibitor of epoxide hydrase) inhibited the formation of all three dihydrodiols of benzo[a]pyrene [755]. In a further study, the same group isolated by HPLC the metabolite benzo[a]pyrene-4,5-epoxide in the presence of the epoxide

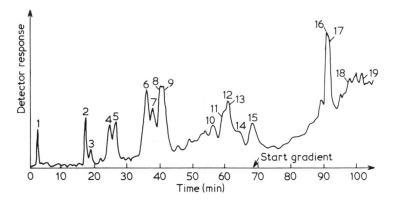

Fig. 16.5 Separation of polynuclear aromatic hydrocarbons in an air sample. Column packing, Zorbax ODS; column dimensions, 250 x 2.1 mm i.d.; eluent, methanol/water (65:35), gradient of 1 per cent methanol min^{-1} initiated after 70 min; pressure, 1200 psi (8.2 MPa); flow rate, 0.21 ml min^{-1}; column temperature, 60° C; detection, UV at 254 nm. Solutes: 1, solvent; 2, phenanthrene; 3, anthracene; 4, fluoranthene; 5, pyrene; 6, triphenylene; 7, benzo[ghi]-fluoranthene; 8, chrysene; 9, benzo[a]anthracene; 10, benzo[j]fluoranthene; 11, benzo[b]fluoranthene; 12, benzo[e]pyrene; 13, perylene; 14, benzo[k]-fluoranthene; 15, benzo[a]pyrene; 16, benzo[ghi]perylene; 17, indeno[123-cd]pyrene; 18, anthanthrene; 19, coronene. Reproduced with permission from reference 760.

hydrase inhibitor [756]. In this case the Permaphase ODS column was buffered at pH 9.0.

In an independent study of the mechanism of formation of benzo[a]pyrene metabolites via oxygenated intermediates, HPLC on two 1 m coupled Permaphase ODS columns was used to investigate the capacity ratios of 24 potential metabolites of benzo[a]pyrene [757]. These included 12 phenols, 5 quinones, 4 dihydrodiols and 3 oxides. The technique was useful for separating the compounds according to type (phenols, quinones etc.) but only partial resolution of individual group members was achieved [757].

Quantitative analysis of benzo[a]pyrene in complex mixtures of aromatic hydrocarbons was carried out by preliminary isolation of benzo[a]pyrene and related molecules on a cross-linked cellulose acetate column. Using fluorimetric detection, 0.8 ng of the compound was detected [758]. Chromatography on Vydac RP has been used for the analysis of benzo[a]pyrene in smoke condensates [759].

HPLC has also been applied to the analysis of polynuclear aromatic hydrocarbons (PAHs) in atmospheric particulate matter. In one study [760], New York City suspended particulate matter was collected on glass fibre filters, extracted with cyclohexane, and the PAH fraction isolated by TLC. Subsequent elution of this fraction on a 250 x 2.1 mm i.d. Zorbax ODS column operated at 60° C with a methanol/water gradient gave a chromatogram with 18 identified

PAH peaks, as shown in Fig. 16.5. A comparison of methods for the quantitation of PAHs by three methods was carried out. In an HPLC/UV method, the UV absorbance of the collected fractions was used to calculate the concentration of each PAH. In a second HPLC method, the peak height of each PAH was compared with that of a standard of known concentration, and in a third method, GLC peak areas were used. The GLC and HPLC peak height methods were in good agreement while the HPLC/UV method, which should be subject to fewer interferences, gave consistently lower results, suggesting that some of the chromatographic peaks included more than one compound. Quantitative analysis using HPLC peak heights suffered from the poorer resolution of the HPLC methodology used while the GLC method suffered from the non-selective response of the flame ionization detector. The advantage of HPLC on reversed phase packings was that isomeric PAHs such as benzo-[a]pyrene and benzo[e]pyrene were readily separated whereas they are not normally separated by GLC [760]. In a recent paper [761], HPLC on Zorbax ODS followed by on-line fluorimetric detection was used to separate 35 and identify 11 PAHs in atmospheric particulate matter and as little as 25 pg of benzo[a]pyrene was detected (signal:noise = 2:1) with this system. Benzo-[a]pyrene and benzo[e]pyrene were separated from each other and from other PAHs by reversed phase chromatography and monitored spectrofluori-metrically [672].

16.3.8 Aromatic hydrocarbons

The analysis of a wide range of aromatic hydrocarbons has been extensively studied and will be mentioned briefly. Reports have appeared on the HPLC analysis of hydrocarbon compounds in the petroleum industry [762—764] and the contribution of HPLC techniques to the on-going American Petroleum Institute Research Project No. 6, which is aimed at separating and identifying the components of commercial petroleum fractions, has been summarized [765]. Using reversed phase chromatography on Permaphase ODS, Sleight-[766] studied the increasing retention of aromatic hydrocarbons with increasing number of aromatic rings and of alkyl substituents, and Vaughan and co-workers [767] used reversed phase chromatography on C_{18}/Corasil for the separation of polynuclear hydrocarbons in used engine oils for forensic purposes. The same workers bonded octadecyltrichlorosilane to Partisil-5 silica (7 μm) and used the resulting phase to study the separations of PAHs [680]. Standard mixtures and engine oil extracts were chromatographed, with fluorimetric detection being used to increase both the selectivity and sensitivity of the HPLC procedure. It was found that using fluorimetric monitoring 10—100 pg of PAHs could be detected after elution. Cigarette smoke condensates [768] and auto exhaust condensates [209] have been chromatographed on Permaphase ODS after size fractionation by exclusion chromatography.

Charge transfer interactions have been used to separate PAHs [95, 769, 770] using for example a Corasil II column impregnated with 2,4,7-trinitrofluorenone [769] and the analysis of a wide range of PAHs by adsorption chromatography on silica [771] and alumina [772] has been reported.

16.4 Industrial pollutants in the environment

16.4.1 Phenols
Several papers have appeared dealing with the analysis of phenols in environmental samples. These include the analysis of phenols in complex mixtures [773] and in industrial wastes [774] using UV detection, and a number of papers dealing with the use of fluorimetric detectors to monitor trace phenols in aqueous environmental samples [484, 775–778]. Chlorophenols, which can be formed during the disinfection of industrial waste by chlorination, were analyzed as quinones after oxidation by ruthenium tetroxide [779].

16.4.2 Nitro compounds
HPLC is suited to the determination of potentially unstable nitro compounds since the compounds can be chromatographed quickly at ambient temperatures. The technique has been applied to the analysis of nitroglycerin in waste waters [780], and in propellants [781, 782], trinitrotoluene in munition waste waters [783], and propellants [782], to the analysis of nitration acids (nitric acid and sulphuric acid) in the industrial nitration of toluene [784], and the analysis of tetryl (N-methyl-N,2,4,6-tetranitroaniline) [785].

16.4.3 Miscellaneous
HPLC has also been applied to the analysis of trimethylolpropane and pentaerythritols in industrial synthesis solutions [786], 2-mercapto-benzothiazole in waste dump effluent [787], polythionates in mining waste water [788] and in Wackenroders solution [789], isocyanates in working atmospheres [790], carcinogenic amines [791], aniline in waste water [103] polyethylene oxide fatty acid surfactants in industrial process waters [792] and phthalate esters in river water [793]. Organomercurials, lead alkyls and other metal ions have also been analyzed by HPLC and are discussed in Chapter 19.

Part V

Miscellaneous applications

17

Miscellaneous plant products

With the improvement in HPLC technology and with the increased efficiencies thus obtained, it has become possible to analyze more complex mixtures previously only tackled by GLC. One field in which there is great interest is the analysis of natural plant products, characteristic of the species being examined. These natural products may be of pharmaceutical interest or may be the colour, flavour, or aroma constituents of commercial products.

Several naturally-occurring xanthones were methylated and separated on a column of MicroPak CN using hexane/chloroform (65:35) as the mobile phase [794]. Moderate success was also obtained in separating the free hydroxy compounds, but a MicroPak NH_2 column with a more polar mobile phase gave superior resolution for the tri- and tetra- hydroxy compounds

Four anthraquinones, chrysophanol, physcion, emodin and aloe emodin, from plant extracts, were separated in less than 20 min on two coupled 3′ x 1/8″ o.d. columns packed with Corasil II [795]. An exponential gradient from hexane to ethyl acetate was used and the compounds were detected at 280 nm. Mixtures of isomeric isoflavones have also been separated by adsorption chromatography on LiChrosorb SI 60 and isomeric biflavonoid compounds were resolved on Pellosil HC [523]. A series of phenolic acids, cinnamic and benzoic acids, and a variety of flavonoid compounds, including flavones, flavonols and flavonones as well as glycosylated flavone derivatives, were separated by reversed phase chromatography on μBondapak/C_{18} using water/acetic acid/methanol mixtures as mobile phases [796]. The minimum detectable amounts were below 50 ng with a UV detector. A pellicular polyamide packing (Pellidon) was found to be useful for the separation of molecules with amino and phenolic groups, retention being determined largely by the ability of the polyamide to form hydrogen bonds with the solutes [523]. This material was found to be useful for the resolution of mixtures of chalcones and flavonones. These species exist in equilibrium as shown in Fig. 17.1 and are often difficult to separate. The O-glycosides of glycoflavones differing only in the position of the hydrolyzable sugar in the flavone skeleton have been separated on columns of LiChrosorb RP8 and LiChrosorb NH_2 [797]. An acetonitrile/water gradient system was used.

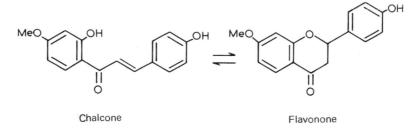

Chalcone   Flavonone

Fig. 17.1   Chalcone–flavonone equilibrium.

Organic acids from a crude extract of maize leaves have been determined by liquid–liquid partition chromatography on Bio-Sil A (20–44 μm) [798]. An aqueous stationary phase was used with a gradient of t-amyl alcohol in chloroform/ cyclohexane (50:50) as the mobile phase.

The quantitative determination of the phytohormone abscisic acid using HPLC has been described by both partition chromatography on Aminex A-6 and ion exchange chromatography on Zipax AAX [799]. The determination of abscisic acid in plant extracts has been performed on a column of Zipax SCX [800]. Using UV detection, 2 ng [799] and 10 ng [800] of abscisic acid could be determined by the two methods. Another group of plant hormones which are structurally related to adenine, cytokinins, have also been analyzed by HPLC [801, 802].

There is a general problem when analyzing a small number of fairly low molecular weight components in a complex matrix such as plant or food extracts, and in many cases preliminary size separation by exclusion chromatography can be used to clean up the sample according to molecular weight prior to the separation of individual components by HPLC. For example, separation of the essential oils from *Lindera umbellata* and *L. sericea* by exclusion chromatography on μStyragel using tetrahydrofuran as solvent gave five peaks [803]. Analysis of the samples on μBondapak/C_{18} with methanol/water (50:50) as mobile phase resolved them into a total of more than 20 peaks. The analysis of cinnamon oils has been performed on Corasil II (37–50 μm) [804]. Quantitative analyses were performed isocratically using an eluent of cyclohexane/ethyl acetate (99:1) and qualitative analyses were carried out using a gradient elution system. A UV detector was used at 260 nm.

Of the two isomeric spiro ethers (Fig. 17.2 I and II) isolated from chamomile, only the *cis* form (I) has anti-spasmodic and anti-inflammatory properties. A quantitative analysis of a mixture of the two isomers was performed by adsorption chromatography on silica with an eluent of hexane/chloroform (60:40) and UV detection at 254 nm [805]. Cis/trans isomerization was observed in chloroform solution but not in cyclohexane.

The fungus *Aspergillus versicolor* produces a large number of metabolites, some of which are known carcinogens and it was thus of interest to develop a method for their detection. Separations were achieved on μPorasil with hexane/

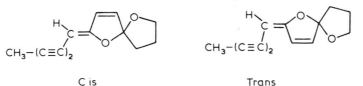

Fig. 17.2   Structure of isomeric spiro ethers.

n-propanol/acetic acid (99.3:0.7:0.1) or hexane/ethyl acetate/acetic acid (83:17:1) as mobile phase or on Partisil-10 PAC with an eluent of hexane/chloroform/acetic acid (65:35:1) [806].

Metabolities of the fungus *Alternaria* were extracted from grain sorghum or grain cultures with methanol and, after a clean-up procedure, analyzed on columns of Corasil II or Zorbax SIL [807]. Mobile phases of either isooctane/tetrahydrofuran or petroleum ether/tetrahydrofuran were used with a system of solvent programming. The metabolites were detected by simultaneous monitoring of fluorescence and absorbance. Quantitation of alternariol monomethyl ether, alternariol and altenuene was based on fluorescence and less than 0.1 ppm of each compound in the grain samples could be detected.

18

Analysis of food products

Many HPLC methods for the analysis of certain compound types in foodstuffs have been mentioned earlier, including sugars (Chapter 11), nucleotides (Chapter 13), vitamins (Chapter 15), xanthines (Section 6.4.2) and pesticides (Chapter 16). Further analyses of nucleotides [808] and sugars [809] in foodstuffs have appeared.

The isolation of flavour fractions from citrus fruits and alcoholic beverages by exclusion chromatography followed by HPLC on Permaphase ETH has been described [810], as has the use of sequential adsorption and reversed phase HPLC in the chromatography of cocoa butter essence [811]. The application of HPLC to the analysis of such flavours as methylanthranilate in grape juice [812], vanillin and ethyl vanillin [813], hop bitter acids [814] and hop resin acids [815] has been reported. HPLC has also been applied to the quality control of flavours [816]. Hydroxy flavones and the dietary xanthines (caffeine, theobromine and theophylline) have been separated from tea by reversed phase chromatography [817]. The analysis of a wide range of compounds in wine, including aromatic phenolic acids, biogenic amines and sugars has been published [818]. Dicarboxylic acids have been estimated in wine and grape must by cation exchange chromatography [819]. Aminobutyric acid and arginine in orange juice [820] and organic acids in foodstuffs [821] have been analyzed by HPLC.

Reversed phase chromatography on ODS-bonded silicas has been used to analyze the artificial sweetener neohesperidin dihydrochalcone [822, 823] and saccharin has been chromatographed by anion exchange chromatography on Zipax SAX [824]. HPLC methods for the analysis of preservatives such as sorbic acid and benzoic acid in a variety of foodstuffs and beverages have been reported [824–827].

The analysis of a range of sulphonic acid dyestuffs of interest as colourings in the foodstuffs and cosmetics industry has been reported [828–834] usually using ion exchange chromatography. The use of reversed phase ion pair partition chromatography was also mentioned earlier for the separation of sulphonic acid dyestuffs [224].

19

Organometallic and inorganic complexes

The analysis of organometallic and inorganic complexes is often complicated by the thermal, photochemical, oxidative or hydrolytic instability of such complexes. Although column [835, 836] and thin layer [837] chromatographic techniques have been widely employed for the separation of inorganic complexes, such techniques require the molecules to be in contact with the adsorbent, usually in the presence of air, for extended periods, and the techniques often lack the required resolution. Gas chromatography has also been used [838] but is often ruled out because of the high operating temperatures involved. HPLC promises to be a useful technique for analytical and preparative scale separations of these complexes since analyses are performed rapidly, degassed solvents are used and light is excluded (using metal columns). Adsorption, liquid—liquid partition and reversed phase chromatography have all been successfully applied to the separation of metal complexes. Ultraviolet detectors have usually been employed to monitor the complexes although polarographic [333], spray impact [82] and atomic absorption [78, 79] detectors have also been used.

An example of the resolution obtainable using HPLC is shown in Fig. 19.1 where four tricarbonyl(phenylcycloheptatriene) iron isomers (differing only in the position of a hydrogen atom in the seven-membered ring) are completely resolved [839]. This separation has already been mentioned (p. 10). Carbonyl and phosphine complexes of ruthenium, iron, iridium and cobalt have been chromatographed by reversed phase HPLC [840]. HPLC separations of diene tricarbonyliron [841] and arene tricarbonylchromium complexes [842] have also been reported. The chromatographic resolution of chiral ferrocenes has been achieved [843, p. 197] and other metallocenes [844] and metallocarboranes [845] have been separated by HPLC. A range of chelate complexes of mercury [846], copper [78, 847–850], nickel [847–850], cobalt [848, 849, 851], palladium [848, 850], platinum [851], beryllium [849], iron [849] and other metals [849] as well as triphenylphosphine complexes of rhodium and iridium [852] have been chromatographed by HPLC. The chromatography of

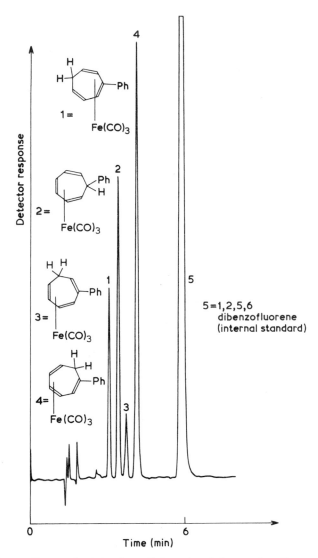

Fig. 19.1 Separation of tricarbonyl(phenylcycloheptatriene)iron isomers. Column packing, Hypersil silica (5 μm); column dimensions, 125 x 5 mm i.d.; eluent, dry hexane; pressure, 220 psi (1.5 MPa); column temperature, 20° C; detection, UV at 254 nm; sensitivity, 0.2 AUFS. Solutes, shown on chromatogram. Reproduced with permission from reference 839.

organomercury [481] and organolead [79] compounds, lead ions [853] and radioactive cadmium ions [854] has also been investigated by HPLC. The polarographic detection of vitamin B_{12} following reversed phase chromatography has been referred to already [333, p. 166].

20

Optical resolution

The resolution of optical enantiomers by liquid chromatography, as in gas chromatography, can be performed by two methods *viz.*, preparation and separation of diastereoisomers using non-chiral mobile and stationary phases or by chromatography of the enantiomers themselves on supports coated or bonded with optically active phases. The two approaches are discussed in turn.

Helmchen and Strubert separated mixtures of racemic amines by HPLC as diastereoisomers after reaction with optically pure O-methyl mandelyl chloride [855]. A 200 x 3 mm i.d. column packed with fractionated LiChrosorb SI 60 5 µm silica was used with a mobile phase of isooctane/ethyl acetate (80:20) to separate the diastereoisomeric amides. The optical purity of the mixture of diastereoisomers was determined by the polarimetric method and compared with the results of quantitation of the chromatograms by the cut and weigh technique. The results of the HPLC method were within the range of precision to be expected by the polarimetric method and quantities down to 0.1 per cent of one enantiomer in the other were measured. Helmchen has also described the preparative separation of diastereoisomeric amides (0.8 g per injection) on highly efficient silica columns which gave reduced plate heights of 2 [856].

The optical purity of the racemic acids (Fig. 20.1, 1–6) was determined by formation of diastereoisomers between the acid chloride and the optically active amine, α-naphthylethylamine [843]. The diastereoisomers from acids (1)–(4) were separated by adsorption chromatography on 20–30 µm silica using hexane/chloroform (97.5:2.5) as mobile phase. The diastereoisomers of acids (5) and (6) were separated by liquid–liquid partition chromatography on Carbowax/Corasil using isooctane/n-butanol (99.5:0.5) as mobile phase. Similarly the estimation of the optical purity of chiral amines was achieved by forming diastereoisomers with optically pure α-phenylbutyric acid followed by HPLC analysis [843].

The amino acids leucine, isoleucine, phenylalanine, and alanine were resolved as their N-d-10-camphorsulphonyl p-nitrobenzyl esters on a column of MicroPak SI-5 with isooctane/isopropanol (98.5:1.5) as eluent [857]. The p-nitrobenzyl

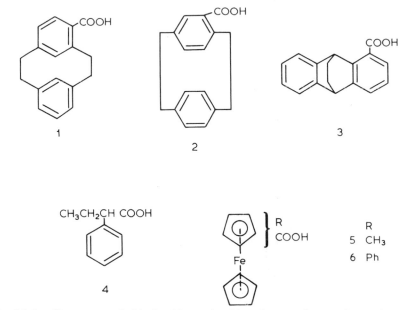

Fig. 20.1 Structures of chiral acids resolved by diastereoisomer formation.

esters were used to improve the detection level at 254 nm. The method was recently extended to include the amino acids methionine, glutamic acid, tryptophan and tyrosine using MicroPak NH_2 (10 μm) columns [858].

Several acyclic isoprenoid acid enantiomers were separated by preparation of amides with (R)-(+)-α-methyl-*p*-nitrobenzylamine and chromatography on silica columns [859, 860]. Using this method citronellic acid samples from natural citronellal and (+)-pulegone were shown to contain respectively 80 per cent and 96–98 per cent excesses of the (3R)-(+)-enantiomer [859].

Some considerable progress has also been made towards the direct resolution of enantiomers using optically active phases. Using silica columns coated with a water/hexafluorophosphate stationary phase and a chloroform solution of optically pure (I) (the chiral complexing agent) as the mobile phase (Fig. 20.2), Cram and coworkers achieved the separation of enantiomeric amines by ion pair partition chromatography [861]. Examples of racemates separated included α-phenylethylamine, phenylglycine, and *p*-hydroxyphenylglycine. Modest pressure drops (< 100 psi; 0.7 MPa) or gravity feed were used and the eluate was monitored by conductivity. Subsequently, covalent attachment of (RR)-I to silica allowed direct resolution of racemic amine (especially amino ester) salts by liquid–solid chromatography [862]. Chloroform or dichloromethane mobile phases were used. Baseline separations were achieved for the enantiomeric pairs of methyl esters of phenylalanine and tryptophan. Covalent attachment of (RR)-II (Fig. 20.3) to a macroreticular cross-linked polystyrene/divinylbenzene resin produced a stable support capable of producing total optical resolution of

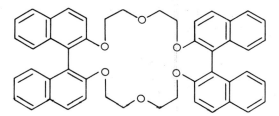

Fig. 20.2 Structure of chiral complexing agent I.

amino acids and ester salts on an analytical or preparative scale [863]. Selectivity factors (α) ranging from 26 to 1.4 were obtained.

Optical resolution of racemic 2,4-dinitrophenyl methyl sulphoxide was achieved on a silica column *in situ* coated with $0.16M$ (R)-(−)-2,2,2,-trifluoro-1-(9-anthryl)ethanol [864]. The mobile phase was a carbon tetrachloride solution of the stationary phase. The yellow sulphoxide formed a red π complex with the carbinol which was seen to move slowly down the column. After recycling the effluent for several hours, material of 68 per cent optical purity was obtained.

Ten racemic helicenes ([5]−[14]) and two double helicenes (diphenanthro [4,3-a; 3′,4′-o] picene and 8,20-dibromodiphenanthro [4,3-a; 4′,3′-j] chrysene) were resolved using high performance liquid chromatography [865]. (R)-(−)- and (S)-(+)-2-(2,4,5,7-tetranitro-9-fluorenylideneaminooxy)propionic acid (TAPA) and three (R)-(−)-homologues derived from butyric (TABA), isovaleric (TAIVA) and hexanoic (TAHA) acids were used as chiral charge-transfer complex-forming stationary phases, *in situ* coated on silica microparticles. A mobile phase of cyclohexane/dichloromethane (75:25) was used. The bulkiness of the group at the chiral centre of the acceptor was found to be critical for the ease of resolution of the donors. The [6]−[14]-helicenes and the double helicenes were completely resolved on (R)-(−)-TAPA. However, the optical isomers of [5]-helicene could be separated only on (R)-(−)-TABA and baseline resolution required ten recycling steps. The helicenes were also resolved on

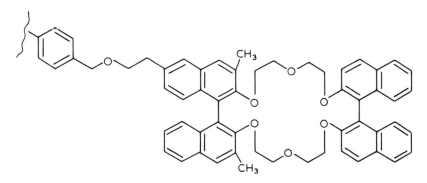

RR−II

Fig. 20.3 Structure of chiral complexing agent II.

TAPA covalently linked to Partisil 7. The values of the resolution factors for [6]-, [7]- and [8]-helicene were found to be even higher than on coated columns.

Grushka and Scott have investigated the potential of using a polyglycine bonded support for the separation of amino acids [866]. Replacement of the polyglycine chain by a peptide chain containing asymmetric centres might produce a bonded phase capable of resolving enantiomeric pairs. The separation of some isomeric dipeptides has been accomplished using bonded optically active tripeptide stationary phases [867]. Although the efficiencies of the columns were poor the separations obtained were superior to those possible on silica gel.

Appendix 1

The Wilke–Chang equation

The Wilke–Chang equation [14], an empirical equation useful for calculating diffusion coefficients for HPLC, is given by

$$D_{\mathrm{m}} = \frac{7.4 \times 10^{-12} \, T\sqrt{(\psi M_{\mathrm{eluent}})}}{\eta V_{\mathrm{solute}}^{0.6}}$$

where D_{m} is the diffusion coefficient of the solute in the eluent ($m^2 \, s^{-1}$); T is the temperature ($^\circ K$); ψ is an eluent association factor (1 for non-polar solvents, 1.5 for ethanol, 1.9 for methanol and 2.6 for water); M_{eluent} is the eluent molecular weight (g); η is the eluent viscosity (in mPa s where $1 \, \text{mPa s} = 1 \, \text{cP} = 1 \, \text{mN s m}^{-2}$); and V_{solute} is the solute molar volume (ml) i.e. the solute molecular weight/density. The equation is accurate to around ±20 per cent.

Some data calculated using the Wilke–Chang equation are given below.

Solvent	Viscosity mPa s	Temperature °C	Solute	Solute molar volume ml	Diffusion coefficient $m^2 s^{-1}$ x 10^9
pentane	0.235	20	toluene	106	4.8
hexane	0.326	20	,,	,,	3.8
,,	0.294	25	,,	,,	4.2
cyclohexane	1.0	20	,,	,,	1.2
chloroform	0.58	,,	,,	,,	2.5
diethyl ether	0.23	,,	,,	,,	4.9
tetrahydrofuran	0.55	,,	,,	,,	2.0
dichloroethane	0.79	,,	,,	,,	1.6
acetonitrile	0.35	,,	,,	,,	2.4
methanol	0.60	,,	,,	,,	1.7
,,	0.40	50	,,	,,	2.8
ethanol	1.2	20	,,	,,	0.91
water	1.00	,,	,,	,,	0.90
,,	0.547	50	,,	,,	1.82
,,	0.282	100	,,	,,	4.1

Solvent	Viscosity mPa s	Temperature °C	Solute	Solute molar volume ml	Diffusion coefficient $m^2 \, s^{-1} \times 10$
hexane	0.326	20	toluene	106	3.8
,,	,,	,,	benzene	89	4.2
			bromobenzene	104	3.8
,,	,,	,,	nitrobenzene	102	3.8
,,	,,	,,	naphthalene	136	3.2
,,	,,	,,	phenanthrene	173	2.8
,,	,,	,,	acetophenone	117	3.5
MeOH/H$_2$O (100:0)	0.60	20	phenol	88	1.9
,, (80:20)	1.20	,,	,,	,,	0.92
,, (60:40)	1.69	,,	,,	,,	0.64
,, (40:60)	1.83	,,	,,	,,	0.58
,, (20:80)	1.59	,,	,,	,,	0.65
,, (0:100)	1.00	,,	,,	,,	0.97

The table is compiled from data in (a) Bristow, P. A. (1976), *Liquid Chromatography in Practice*, HETP, Macclesfield, UK, p. 250 and (b) Bristow, P. A. and Knox, J. H. (1977) *Chromatographia*, 10, 279.

Appendix 2

Electronic absorption bands for representative chromophores

Chromophore	System	λ Max	ε Max	λ Max	ε Max	λ Max	ε Max
Ether	−O−	185	1000				
Thioether	−S−	194	4600	215	1600		
Amine	−NH₂	195	2800				
Thiol	−SH	195	1400				
Disulphide	−S−S−	194	5500	255	400		
Bromide	−Br	208	300				
Iodide	−I	260	400				
Nitrile	−C≡N	160	−				
Acetylide	−C≡C−	175−180	6000				
Sulphone	−SO₂−	180	−				
Oxime	−NOH	190	5000				
Azido	>C=N−	190	5000				
Ethylene	−C≡C−	190	8000				
Ketone	>C=O	195	1000	270−285	18−30		
Thioketone	>C=S	205	strong				
Esters	−COOR	205	50				
Aldehyde	−CHO	210	strong	280−300	11−18		
Carboxyl	−COOH	200 210	50 70				
Sulphoxide	>S → O	210	1500				
Nitro	−NO₂	210	strong				
Nitrite	−ONO	220−230	1000−2000	300−4000	10		
Azo	−N=N−	285−400	3−25				
Nitroso	−N=O	302	100				
Nitrate	−ONO₂	270 (shoulder)	12				
	−(C=C)₂− (acyclic)	210 230	21,000				
	−(C=C)₃−	260	35,000				
	−(C=C)₄−	300	52,000				
	−(C=C)₅−	330	118,000				
	−(C=C)₂− (alicyclic)	230−260	3000−8000				
	C=C−C≡C	219	6,500				
	C=C−C≡N	220	23,000				
	C=C−C=O	210−250	10,000−20,000			300−350	weak
	C=C−NO₂	229	9,500				
Benzene		184	46,700	202	6,900	255	170
Diphenyl				246	20,000		
Naphthalene		220	112,000	275	5,600	312	175
Anthracene		252	199,000	375	7,900		
Pyridine		174	80,000	195	6,000	251	1,700
Quinoline		227	37,000	270	3,600	314	2,750
Isoquinoline		218	80,000	266	4,000	317	3,500

Reproduced with permission from Willard H. H., Merritt Jr, L. L. and Dean J. A. (4th ed) (1965). *Instrumental Methods of Analysis*, D. Van Nostrand Co., Inc., New York.

Appendix 3

Packing materials for adsorption and liquid–liquid partition chromatography

Type	Designation	Particle size μm	Surface area m² g⁻¹	Shape[a]	Supplier
Pellicular silica	Corasil I[b]	37–50	7	S	Waters
	Corasil II[b]	37–50	14	S	Waters
	Liqua-Chrom	44–53	<10	S	Applied Science
	Pellosil HC[c]	37–44	8	S	Whatman
	Pellosil HS[c]	37–44	4	S	Whatman
	Perisorb A	30–40	14	S	Merck[d]
	SIL–X–II	30–40	12	S	Perkin-Elmer
	Vydac adsorbent[e]	30–44	12	S	Separations Group
	Zipax	25–37	~1	S	Du Pont
Porous silica	Bio-Sil A	20–44	>200	I	BioRad
	Davidson Code 62	>150	350	I	W. R. Grace
	Dia-Chrom	37–44	10	I	Applied Science
	Hypersil	5	200	S	Shandon Southern
	LiChrospher SI 100	10	25	S	Merck[d]
	LiChrosorb[f] SI 60	5, 10, 30	500	I	Merck[d]
	LiChrosorb[f] SI 100	5, 10, 30	400	I	Merck[d]
	Nucleosil	5–10	–	S	Macherey-Nagel
	MicroPak SI 5, SI 10	5, 10	500	I	Varian
	Partisil 5, 10, 20	5, 10, 20	400	I	Whatman
	μ Porasil	10	400	I	Waters
	Porasil A	37–75 or 75–125	350–500	S	Waters
	Porasil B	„	125–250	S	Waters
	Porasil C	„	50–100	S	Waters
	Porasil D	„	25–45	S	Waters
	Porasil E	„	10–20	S	Waters
	Porasil F	„	2–6	S	Waters
	Porasil T	15–25	300	I	Waters
	SIL-X	32–45	300	I	Perkin-Elmer
	SIL-X-I	13	400	I	Perkin-Elmer
	Spherisorb[g] S5W, S10W, S20W	5, 10, 20	200	S	Phase Separations; Spectra Physics

vpe	Designation	Particle size	Surface area	Shape[a]	Supplier
	Spherosil XOA-400	Choice of: <40 ... 100—150	350—500	I	Rhone-Progil
	XOA-200	,,	125—250	S	Rhone-Progil
	XOA-075	,,	50—100	S	Rhone-Progil
	XOB-030	,,	25—45	S	Rhone-Progil
	XOB-015	,,	10—20	S	Rhone-Progil
	XOB-005	,,	2—6	S	Rhone-Progil
	XOA-800	5	800	S	Rhone-Progil
	XOA-600	5	600	S	Rhone-Progil
	Vydac TP silica[h]	10	200	I	Separations Group
	Zorbax SIL	5	300	S	Du Pont
ellicular alumina	Pellumina HC[c]	37—44	8	S	Whatman
	Pellumina HS[c]	37—44	4	S	Whatman
orous alumina	Bio-Rad AG LiChrosorb[f]	<75	>200	I	BioRad
	Alox T	5, 10, 30	70	I	Merck[d]
	MicroPak Al-5, Al-10 Spherisorb[g]	5, 10	70	I	Varian
	A5W, A10W, A20W	5, 10, 20	95	S	Phase Separations; Spectra Physics
	Woelm Alumina	18—30	>200	I	Woelm

a   S = spherical; I = irregular.
b   Corasil II has a thicker silica coating than Corasil I.
c   HC (high capacity) material has a thicker coating than the HS (high speed) material.
d   E. M. Labs in the USA.
e   Chemically deactivated; avoid solvents more polar than methanol.
f   Formerly marketed as Merckosorb.
g   Developed by the Atomic Energy Research Establishment (A.E.R.E.), Harwell, UK.
h   TP = totally porous.
Some representative average pore diameters of the porous silicas are Hypersil (10 nm), LiChrosorb SI60 (6 nm), LiChrosorb SI100 (10 nm), Partisil 5, 10, 20 (5.5—6.0 nm), Porasil A (<10 nm), Porasil B (15 nm), Porasil F (>150 nm), Spherisorb S5W and S10W (8 nm), Spherosil XOA 400 (8 nm), Spherosil XOA 200 (15 nm), Spherosil XOB 015 (125 nm) and Zorbax SIL (7.5 nm).

Appendix 4

Polymer resins and polymer-coated materials for HPLC

Use	Type	Designation	Particle size μm	Approx. exchange capacity μequiv. g⁻¹	Supplier
Liquid–liquid partition	Polymer-coated glass beads (or superficially porous glass beads)	Pellidon[a]	45	n.a.[e]	Whatman
		Perisorb PA6[b]	30–40	n.a.[e]	Merck[f]
		Zipax-ANH[c]	37–44	n.a.[e]	Du Pont
		Zipax-HCP[d]	25–37	n.a.[e]	Du Pont
		Zipax-PAM[a]	25–37	n.a.[e]	Du Pont
Anion exchange	Polymer-coated glass beads (or superficially porous glass beads)	AE-Pellionex-SAX[g]	44–53	10	Whatman
		AL-Pellionex-WAX[h]	44–53	–	Whatman
		AS-Pellionex-SAX[g]	44–53	10	Whatman
		Pellicular Anion[g]	40	10	Varian
		Zipax-SAX[g]	25–37	12	Du Pont
		Zipax-WAX[h]	25–37	12	Du Pont
Cation exchange[i]	Polymer-coated glass beads (or superficially porous glass beads)	HC-Pellionex-SCX	44–53	60	Whatman
		HS-Pellionex-SCX	44–53	10	Whatman
		Pellicular cation	40	10	Varian
		Zipax SCX	25–37	5	Du Pont

Anion exchange[g]	Porous polymer resins[j]	Aminex A-28	7–11	3200	Bio-Rad
		DA-X8A	7–9	4000	Durrum
		DA-X4	15–25	2000	Durrum
		HA-X4 (X6, X8, X10)	7–10; 15–25	n.d.[k]	Hamilton
Cation exchange[i]	Porous polymer resins[j]	Aminex A-7	7–11	5000	Bio-Rad
		B-80	10–20	5200	Hamilton
		HC-X8.00	7–10	n.d.[k]	Hamilton
		PA-35	13	5000	Beckman
		DC-4A	7–9	5000	Durrum

a Contains a nylon coating.
b Contains a polycaprolactam coating.
c Contains a cyanoethylsilicone coating.
d Contains a saturated hydrocarbon polymer coating.
e n.a. = not applicable.
f E.M. Labs in the USA.
g Contain $-NR_3^+$ functional groups (strong anion exchangers).
h Contain $-NH_2$ functional groups (weak anion exchangers).
i Contain $-SO_3^-$ functional groups (strong cation exchangers).
j Usually available in a variety of size fractions.
k n.d. = not determined.

Appendix 5
Chemically bonded stationary phases for HPLC

Bonded phase	Designation	Support type	Particle size μm	Supplier
Octadecyl hydrocarbon ($C_{18}H_{37}Si$-)	Bondapak C_{18}/Corasil	Pellicular	37–50	Waters
	Co:Pell ODS	Pellicular	41	Whatman
	ODS-SIL-X-II	Pellicular	30–40	Perkin-Elmer
	Perisorb RP	Pellicular	30–40	Merck[a]
	Permaphase ODS	Pellicular	25–37	Du Pont
	Vydac RP	Pellicular	30–44	Separations Group
	Bondapak C_{18}/Porasil	Porous	37–75	Waters
	μBondapak C_{18}/Porasil	Porous	10	Waters
	ODS-Hypersil	Porous	5	Shandon Southern
	LiChrosorb RP-18[b]	Porous	5, 10	Merck
	MicroPak CH[b]	Porous	10	Varian
	Nucleosil C_{18}	Porous	5, 10	Macherey-Nagel
	Partisil-10 ODS[b]	Porous	10	Whatman
	Partisil-10 ODS-II[b]	Porous	10	Whatman
	ODS-SIL-X-I	Porous	8–18	Perkin-Elmer
	Spherisorb S5 ODS (S10 ODS)	Porous	5, 10	Phase Separations; Spectra Physics
	Vydac TP Reversed Phase	Porous	10	Separations Group
	Zorbax ODS	Porous	5	Du Pont
Short Chain Hydrocarbon	LiChrosorb RP-8[c]	Porous	5, 10	Merck
	LiChrosorb RP-2[d]	Porous	5, 10, 30	Merck
	Nucleosil C_8[c]	Porous	5, 10	Macherey-Nagel
	SAS-Hypersil	Porous	5	Shandon Southern
Phenyl	Bondapak Phenyl/Corasil	Pellicular	37–50	Waters
	Bondapak Phenyl/Porasil	Porous	37–75	Waters
	Phenyl-SIL-X-I	Porous	8–18	Perkin-Elmer
Allylphenyl	Allylphenyl-SIL-X-I	Porous	8–18	Perkin-Elmer
Fluoroether	FE-SIL-X-I	Porous	8–18	Perkin-Elmer
Ether	Permaphase ETH	Pellicular	25–37	Du Pont
Nitro	Nucleosil NO_2	Porous	5, 10	Macherey-Nagel

Bonded phase	Designation	Support type	Particle size μm	Supplier
Cyano	Co:Pell PAC	Pellicular	41	Whatman
	Vydac Polar Phase	Pellicular	30–44	Separations Group
	μBondapak CN	Porous	10	Waters
	Cyano-SIL-X-I	Porous	8–18	Perkin-Elmer
	Durapak OPN/Porasil[e]	Porous	37–75	Waters
	Nucleosil CN	Porous	5, 10	Macherey-Nagel
	MicroPak CN	Porous	10	Varian
	Partisil-10-PAC	Porous	10	Whatman
	Spherisorb S5 CN	Porous	5	Phase Separations; Spectra Physics
	Vydac TP Polar Phase	Porous	10	Separations Group
Hydroxyl	Durapak Carbowax 400/Corasil[e]	Pellicular	37–50	Waters
	Durapak Carbowax 400/Porasil[e]	Porous	37–75	Waters
	LiChrosorb Diol	Porous	10	Merck
Amino[f]	APS-Hypersil	Porous	5	Shandon Southern
	Amino-SIL-X-I	Porous	8–18	Perkin-Elmer
	μBondapak NH$_2$	Porous	10	Waters
	LiChrosorb NH$_2$	Porous	10	Merck
	MicroPak NH$_2$	Porous	10	Varian
	Nucleosil NH$_2$	Porous	5, 10	Macherey-Nagel
	Nucleosil N(CH$_3$)$_2$	Porous	5, 10	Macherey-Nagel
Anion Exchanger[g]	Bondapak AX/Corasil	Pellicular	37–50	Waters
	Perisorb AN	Pellicular	30–40	Merck
	Permaphase AAX	Pellicular	25–37	Du Pont
	Partisil-10 SAX	Porous	10	Whatman
	Vydac TP Anion Exchange	Porous	10	Separations Group
Cation Exchanger[h]	Bondapak CX/Corasil	Pellicular	37–50	Waters
	Perisorb KAT	Pellicular	30–40	Merck
	Partisil-10 SCX	Porous	10	Whatman
	Vydac TP Cation Exchange	Porous	10	Separations Group

a E.M. Labs in the USA.
b Carbon loadings are LiChrosorb RP-18 (22 per cent), MicroPak CH (22 per cent),
   Partisil-10 ODS (5 per cent) and Partisil-10 ODS-II (16 per cent)
c Contains bonded octyl groups.
d Support treated with dimethyldichlorosilane.
e Durapak supports are silicate esters i.e. formed via Si—O—C linkages.
f Function as polar bonded phases or weak anion exchangers.
g Contain bonded quaternary ammonium groups.
h Contain bonded sulphonic acid groups.
Note. Waters also market a bonded phase (10 μm) called μBondapak/Carbohydrate for the
   separation of carbohydrates.

Appendix 6

Calculation of surface coverage of a bonded phase

Given that the bonded phase formed by reaction of silica (surface area $200 \, \text{m}^2 \text{g}^{-1}$) with trimethylchlorosilane had a carbon content of 3.0 per cent by elemental analysis, calculate the surface coverage of bonded phase. (Assume that there are five silanol groups per $1 \, \text{nm}^2$.)

Number of silanol groups per gram of silica is given by:

$$(200 \times 10^{18} \, \text{nm}^2 \, \text{g}^{-1}) \times (5 \text{ silanol groups nm}^{-2})$$

$$= 10^{21} \text{ silanol groups per gram silica}$$

$$= \frac{10^{21}}{6.02 \times 10^{23}} \text{ moles silanol groups per gram silica}$$

$$= 1.66 \text{ mmoles silanol groups per gram silica}$$

Carbon content of 3.0 per cent corresponds to 0.03 g carbon per gram of support

$$= \frac{0.03}{12 \times 3} \text{ moles trimethylsilyl groups per gram silica}$$

$$= 0.83 \text{ mmoles trimethylsilyl groups per gram silica}$$

Thus surface coverage $= \dfrac{0.83}{1.66} \times 100\%$

$$= 50\%$$

As shown by Unger et al. [154] about 50 per cent of the total silanol groups can be capped with trimethylsilyl groups. With bulkier molecules, surface coverage is less extensive.

N.B. For the given silica, 0.83 mmoles trimethylsilyl groups per gram

$$= 4.2 \; \mu\text{moles trimethylsilyl groups per m}^2$$

(For di- and trifunctional chlorosilanes a factor must be introduced into the calculation to account for the fact that one molecule can react with more than one silanol groups).

Appendix 7

Use of reversed phase HPLC to measure partition coefficients and to predict biological activity

In correlating the structure of molecules with their biological activity, Hansch [868, 869] showed that successful drug and pesticide molecules possess an optimum hydrophilic/lipophilic balance. Molecules which are too polar do not penetrate lipid barriers and thus fail to reach their site of action; polar molecules are also rapidly excreted from the target organism. Molecules which are too lipophilic are immobilized on the first lipophilic material which they encounter and are thus also prevented from reaching their targets. Hansch has shown [868, 869] that biological activity (measured as the concentration of compound producing a standard response in a constant time interval) correlates much better with the hydrophobic character of a series of test molecules than with their electronic or steric properties.

The normal measure of hydrophobicity is the octanol/water partition coefficient of the neutral molecule; these coefficients have been successfully correlated with biological activity in structure/activity studies [868, 869]. The Environmental Protection Agency also requires octanol/water partition coefficients of candidate pesticides to decide whether bioaccumulation of the compound is likely [870]. For partition coefficients of > 100, further bioconcentration experiments with fish are recommended; for partition values of < 10, bioconcentration is considered unlikely. The traditional method for determining octanol/water partition coefficients by shaking and analysis of the equilibrium concentration of solute in each phase is tedious to perform and requires the compound to be in a high state of purity. It has been demonstrated that k' values of solutes in reversed phase HPLC can be correlated with their octanol/water partition coefficients as well as directly with their biological activity. An advantage of this approach (shared by the related reversed phase TLC method) is that solute purity is less critical than with the traditional method. In a typical reversed phase system using an octadecyl-silica column and a methanol/water eluent, the more hydrophobic a solute, the longer it will be retained (see p. 45). Linear correlations of k' values with traditional partition coefficients using the reversed phase HPLC method have been obtained with hypotensive triaminopyrimidine 3-oxides [871], standard organic compounds (pyridine, acetanilide etc.) [872], nitrosoureas [873], sulphonamides and barbiturates [874], anilines and phenols [875] and chlorinated insecticides [695] (p. 175). Huber and coworkers have also used reversed phase and normal phase liquid—liquid partition chromatography to determine partition coefficients of pesticides and steroids [439]. In addition, k' values obtained by reversed phase HPLC have been correlated with the biological activity of polyene antifungal compounds [265], 1,3,5-triazine herbicides [876] anilines and phenols [875] and sulphonamides and barbiturates [874].

It is probable that reversed phase packings in which residual silanol groups have been reacted with trimethylsilyl groups will give better correlations between k' and the octanol/water partition values, since the HPLC retention will then be by a strictly partition mechanism. McCall [871] observed an improved correlation when the residual silanol groups of C_{18}/Corasil were silylated. As pointed out by Taylor and coworkers the availability of adsorption-free reversed phase supports for HPLC gives this method an advantage over the TLC method using coated silica plates, where silanol groups can interfere [872]; a further advantage of the HPLC method is the wide range of k' values which can be accurately determined [872]. In a recent paper, Horvath and Melander discussed the mechanism of solute retention by reversed phase chromatography and indicated how reversed phase HPLC could be used for the measurement of physical chemical properties [575].

Thus although neither the traditional octanol/water system nor the ODS-silica/aqueous methanol HPLC system is structurally similar to a biological membrane, both systems can be used to predict the transport of molecules across biological tissues. The simplicity of the HPLC method suggests it will be increasingly used in structure activity correlations.

Appendix 8

Packing materials for exclusion chromatography[a]

Designation	Type	Exclusion limit[b] (molecular weight)	Supplier
	Rigid packings		
Bio-Glas	Porous Glass	5×10^6	Bio-Rad
CPG-10[c]	Porous Glass	1.2×10^6	Electro-Nucleonics
Vit-X	Porous Glass	1.2×10^6	Perkin-Elmer
μBondagel E[d]	Porous Silica	2×10^6	Waters
LiChrospher	Porous Silica	$>3.5 \times 10^6$	Merck[e]
Porasil[f]	Porous Silica	5×10^6	Waters
Spherosil	Porous Silica	5×10^7	Rhone-Poulenc
	Semi-rigid packings		
Hydrogel	PS/DVB[g] rendered hydrophilic	2×10^6	Waters
Merckogel-OR	Polyvinylacetate	10^6	Merck
Styragel[h]	PS/DVB[g]	5×10^7	Waters

a   Soft gels such as Sephadex (Pharmacia) and Bio-Gel (Bio-Rad) are not included in the Table as they cannot be used at high pressures.
b   Packings are usually available with a choice of average pore diameters, e.g. Bio-Glas, 20–250 nm; CPG-10, 7.5–200 nm; LiChrospher, 10–400 nm; and Spherosil, 8–300 nm. Exclusion limits refer to the fraction of highest pore diameter.
c   CPG = Controlled Pore Glass; also available (from Corning) with a bonded hydrophilic group (as Glycophase – CPG) which reduces adsorptive effects in exclusion chromatography with aqueous eluents (e.g. of proteins).
d   Contain bonded organosilane groups with an ether functionality.
e   E.M. Labs in the USA.
f   Chemically deactivated.
g   PS/DVB = polystyrene/divinylbenzene.
h   A 10 μm fraction (μStyragel) of Styragel (<37 μm or 37–75 μm) is available in packed columns.

Appendix 9

Comparison of mesh numbers and aperture sizes

ASTM sieve no.	Aperture (μm)
60	250
70	210
80	177
100	150
120	125
150	105
200	74
230	63
270	53
325	44
400	37

References

1. Tswett, M. (1903), *Proc. Warsaw Soc. Nat. Sci. (Biol).*, **14**, No. 6.
2. Kuhn, R. and Lederer, E. (1931), *Naturwiss*, **19**, 306.
3. Reichstein, T. and Van Euw, J., (1938), *Helv. Chim. Acta.*, **21**, 1197.
4. Martin, A. J. P. and Synge, R. L. M. (1941), *Biochem. J.*, **35**, 1358.
5. James, A. T. and Martin, A. J. P. (1952), *Biochem. J.*, **50**, 679.
6. Hamilton, P. B., Bogue, D. C. and Anderson, R. A. (1960), *Anal. Chem.*, **32**, 1782.
7. Snyder, L. R. (1968), *Principles of Adsorption Chromatography*, Marcel Dekker, New York.
8. Giddings, J. C. (1965), *Dynamics of Chromatography. Part I. Principles and Theory*, Marcel Dekker, New York.
9. Kirkland, J. J. (1969), *J. Chromatogr. Sci.*, **7**, 7.
10. Huber, J. F. K. (1969), *J. Chromatogr. Sci.*, **7**, 85.
11. Horvath, C. G., Preiss, B. A. and Lipsky, S. R. (1967), *Anal. Chem.*, **39**, 1422.
12. Halasz, I. and Sebestian, I. (1969), *Angew. Chem. Int. Ed. Engl.*, **8**, 453.
13. Brust, O-E., Sebestian, I. and Halasz, I. (1973), *J. Chromatogr.*, **83**, 15.
14 Wilke, C. R. and Chang, D. (1955), *Amer. Inst. Chem. Eng. J.*, **1**, 264.
15. Van Deemter, J. J., Zuiderweg, F. J. and Klinkenberg, A. A. (1956) *Chem. Eng. Sci.*, **5**, 271.
16. Done, J. N., Kennedy, G. J. and Knox, J. H. (1973), *Gas Chromatography 1972*, (ed. S. G. Perry) Applied Science Publ., London, p. 145.
17. Grushka, E., Snyder, L. R. and Knox, J. H. (1975), *J. Chromatogr. Sci.*, **13**, 25.
18. Endele, R., Halasz, I. and Unger, K. (1974), *J. Chromatogr.*, **99**, 377.
19. Majors, R. E. (1973), *J. Chromatogr. Sci.*, **11**, 88.
20. Kraak, J. C. and Huber, J. F. K. (1974), *J. Chromatogr.*, **102**, 333.
21. Done, J. N. (1976), *Practical High Performance Liquid Chromatography*, (ed. C. F. Simpson), Heyden, London, p. 69.
22. McNair, H. M. and Chandler, C. D. (1974), *J. Chromatogr. Sci.*, **12**, 425.
23. McNair, H. M. and Chandler, C. D. (1976), *J. Chromatogr. Sci.*, **14**, 477.
24. Parris, N. A. (1976), *Instrumental Liquid Chromatography*, Elsevier, Amsterdam, Chapter 4.
25. Henry, R. A. (1971), *Modern Practice of Liquid Chromatography* (ed. J. J. Kirkland), Wiley-Interscience, New York, Chapter 2.
26. Berry, L. and Karger, B. L. (1973), *Anal Chem.*, **45**, 819A.

27. Abbott, S. R., Berg, J. R., Achener, P. and Stevenson, R. L. (1976), *J. Chromatogr.*, **126**, 421.
28. Byrne, S. H., Schmit, J. A. and Johnson, P. E. (1971), *J. Chromatogr. Sci.*, **9**, 592.
29. Henry, R. A. Du Pont LC Technical Bulletin No. 73–1.
30. Jackson, M. T. and Henry, R. A. (1974), *Int. Lab.*, 57.
31. Martin, M., Blu, G., Eon, C., and Guiochon, G. (1975), *J. Chromatogr.*, **112**, 399.
32. Martin, M., Guiochon, G., Blu, G. and Eon, C. (1977), *J. Chromatogr.*, **130**, 458.
33. Achener, P., Abbott, S. R. and Stevenson, R. L. (1977), *J. Chromatogr.*, **130**, 29.
34. Pryde, A. and Darby, F. J. (1975), *J. Chromatogr.*, **115**, 107.
35. Scott, R. P. W. and Kucera, P. (1973), *J. Chromatogr. Sci.*, **11**, 83.
36. Byrne, Jr., S. H. (1971), *Modern Practice of Liquid Chromatography*, (ed. J. J. Kirkland), Wiley-Interscience, New York, Chapter 3.
37. Maggs, R. J., (1976), *Practical High Performance Liquid Chromatography*, (ed. C. F. Simpson), Heyden, London, p. 269.
38. Parris, N. A. (1976), *Instrumental Liquid Chromatography*, Elsevier, Amsterdam, Chapter 6.
39. Polesuk, J. and Howery, D. G. (1973), *J. Chromatogr. Sci.*, **11**, 226.
40. Munk, M. N. (1970), *J. Chromatogr. Sci.*, **8**, 491.
41. Huber, J. F. K. (1969), *J. Chromatogr. Sci.*, **7**, 172.
42. Baker, D. R., Williams, R. C. and Steichen, J. C. (1974), *J. Chromatogr. Sci.*, **12**, 499.
43. Carr, C. D. (1974), *Anal. Chem.*, **46**, 743.
44. Brinkman, U. A. Th., Seetz, J. W. F. L. and Reymer, H. G. M. (1976), *J. Chromatogr.*, **116**, 353.
45. Sparacino, C. M. and Hines, J. W. (1976), *J. Chromatogr. Sci.*, **14**, 549.
46. Smith, R. N. and Zetlein, M. (1977), *J. Chromatogr.*, **130**, 314.
47. Okuyama, S., Vemura, D. and Hirata, Y. (1976), *Chem. Lett.* 679.
48. Lawrence, J. F. and Frei, R. W. (1974), *J. Chromatogr.*, **98**, 253.
49. Lawrence, J. F. and Frei, R. W. (1976), *Chemical Derivatization in Liquid Chromatography*, Elsevier, Amsterdam.
50. Muusze, R. G. and Huber, J. F. K. (1974), *J. Chromatogr. Sci.* **12**, 779.
51. Koen, J. G., Huber, J. F. K., Poppe, H. and Den Boef, G. (1970), *J. Chromatogr. Sci.*, **8**, 192.
52. Kissinger, P. T., Refshauge, C., Dreiling, R. and Adams, R. N. (1973), *Anal. Lett.*, **6**, 465.
53. Fleet, B. and Little, C. J. (1974), *J. Chromatogr. Sci.*, **12**, 747.
54. Knox, J. H., Laird, G. R. and Raven, P. A. (1976), *J. Chromatogr.*, **122**, 129.
55. Vespalec, R. (1975), *J. Chromatogr.*, **108**, 243.
56. Aue, W. A. and Kapila, S. (1973), *J. Chromatogr. Sci.*, **11**, 255.
57. Applications Literature, Pye Unicam, York St, Cambridge, G.B.
58. Dolphin, R. J., Willmott, F. W., Mills, A. D. and Hoogeveen, L. P. J. (1976), *J. Chromatogr.*, **122**, 259.
59. Willmott, F. W. and Dolphin, R. J. (1974), *J. Chromatogr. Sci.*, **12**, 695.
60. Snyder, L. R. (1976), *J. Chromatogr.*, **125**, 287.
61. Spackman, D. H., Stein, W. H. and Moore, S. (1958), *Anal. Chem.*, **30**, 1190.
62. Deelder, R. S. and Hendricks, P. J. H. (1973), *J. Chromatogr.* **83**, 343.

63. Mrochek, J. E., Dinsmore, S. R. and Waalkes, T. P. (1975), *Clin. Chem.*, **21**, 1314.
64. James, A. T., Ravenhill, J. R. and Scott, R. P. W. (1964), *Gas Chromatography* (Fifth Int. Symposium) Brighton, England, (ed. A. Gouldup), Institute of Petroleum, London p. 197.
65. Scott, R. P. W. (1965), British Patent 998,107; (1966), U.S. Patent 3,292,420.
66. Szakasits, J. J. and Robinson, R. E. (1974), *Anal. Chem.*, **46**, 1648.
67. Scott, R. P. W., Scott, C. G., Munroe, M. and Hess, Jr., J. (1974), *J. Chromatogr.*, **99**, 395.
68. Horning, E. C., Carroll, D. I., Dzidic, I., Haegele, K. D., Horning, M. G. and Stillwell, R. N. (1974), *J. Chromatogr.*, **99**, 13.
69. McLafferty, F. W., Knutti, R., Venkataraghavan, R., Arpino, P. J. and Dawkins, B. G. (1975), *Anal. Chem.*, **47**, 1503.
70. Arpino, P. J., Dawkins, B. G. and McLafferty, F. W. (1974), *J. Chromatogr. Sci.*, **12**, 574.
71. Schulten, H-R. and Beckey, H. D. (1973), *J. Chromatogr.*, **83**, 315.
72. Jones, P. R. and Yang, S. K. (1975), *Anal. Chem.*, **47**, 1000.
73. McFadden, W. H., Schwartz, H. L. and Evans, S. (1976), *J. Chromatogr.*, **122**, 389.
74. Stolyhwo, A., Privett, O. S. and Erdahl, W. L. (1973), *J. Chromatogr. Sci.*, **11**, 263.
75. Publication 019/6 250 6/76, Applied Chromatography Systems Ltd., Luton, England.
76. Wilks Scientific Corp., Box 449, S. Norwalk, Ct 06856 USA.
77. Lemar, M., Versaud, P. and Porthault, M. (1977), *J. Chromatogr.*, **132**, 295.
78. Jones IV, D. R. and Manahan, S. E. (1976), *Anal. Chem.*, **48**, 502.
79. Botre, C., Cacace, F. and Cozzani, R. (1976), *Anal. Lett.*, **9**, 825.
80. Freed, D. J, (1975), *Anal. Chem.*, **47**, 186.
81. Davenport, T. B. (1969), *J. Chromatogr.*, **42**, 219.
82. Mowery Jr., R. A. and Juvet Jr., R. S. (1974), *J. Chromatogr. Sci.*, **12**, 687.
83. Julin, B. G., Vandenborn, H. W. and Kirkland, J. J. (1975), *J. Chromatogr.*, **112**, 443.
84. Steichen, J. C. (1975), *J. Chromatogr.*, **104**, 39.
85. Sieswerda, G. B., Poppe, H. and Huber, J. F. K. (1975), *Anal. Chim. Acta.*, **78**, 343.
86. Horwitz, E. P., Delphin, W. H., Bloomquist, C. A. A. and Vandegrift, G. F. (1976), *J. Chromatogr.*, **125**, 203.
87. Knox, J. H. and Parcher, J. F. (1969), *Anal. Chem.*, **41.**, 1599.
88. Knox, J. H. (1976), *Practical High Performance Liquid Chromatography*, (ed. C. F. Simpson), Heyden, London, p. 19.
89. Kraak, J. C., Poppe, H. and Smedes, F. (1976), *J. Chromatogr.*, **122**, 147.
90. Scott, R. P. W. and Kucera, P. (1976), *J. Chromatogr.* **125**, 251.
91. Martin, M., Verillon, F., Eon, C. and Guiochon, G. (1976), *J. Chromatogr.*, **125**, 17.
92. Henry, R. A., Byrne, S. H. and Hudson, D. R., (1974), *J. Chromatogr. Sci.*, **12**, 197.
93. Webber, T. J. N. and McKerrell, E. H. (1976), *J. Chromatogr.*, **122**, 243.
94. Coq. B., Gonnet, C. and Rocca, J.-L. (1975), *J. Chromatogr.*, **106**, 249.
95. Knox, J. H. (1975), *Chem. Ind.*, 29.

96. Cox, G. B., Loscombe, C. R., Slucutt, M. J., Sugden, K. and Upfield, J. A. (1976), *J. Chromatogr.*, **117**, 269.
97. Caude, M., Lefevre, J.-P. and Rosset, R. (1975), *Chromatographia*, **8**, 217.
98. Snyder, L. R. (1970), *J. Chromatogr. Sci.*, **8**, 692.
99. Engelhardt, H. and Elgass, H. (1975), *J. Chromatogr.*, **112**, 415.
100. DeStefano, J. J. and Beachell, H. C. (1972), *J. Chromatogr. Sci.*, **10**, 654.
101. Bombaugh, K. J. and Almquist, P. W. (1975), *Chromatographia*, **8**, 109.
102. Majors, R. E. (1973), *Anal. Chem.*, **45**, 759.
103. Kirkland, J. J. (1974), *Analyst.*, **99**, 859.
104. Bakalyar, S. R. and Henry, R. A. (1976), *J. Chromatogr.*, **126**, 327.
105. Scott, R. P. W. and Reese, C. E. (1977), *J. Chromatogr.*, **138**, 283.
106. Fitzpatrick, F. A., Siggia, S. and Dingman Jr., J. (1972), *Anal. Chem.*, **44**, 2211.
107. Henry, R., A. Schmit, J. A. and Dieckman, J. F. (1971), *J. Chromatogr. Sci.*, **9**, 513.
108. Fitzpatrick, F. and Siggia, S. (1973), *Anal. Chem.*, **45**, 2310.
109. Borch, R. F. (1975), *Anal. Chem.*, **47**, 2437.
110. Cooper, M. J. and Anders, M. W. (1974), *Anal. Chem.*, **46**, 1849.
111. Durst, H. D., Milano, M., Kikta Jr., E. J., Connelly, S. A. and Grushka, E. (1975), *Anal. Chem.*, **47**, 1797.
112. Pei, P. T. S., Kossa, W. C., Ramachandran, S. and Henly, R. S. (1976), *Lipids*, **11**, 814.
113. Zimmerman, C. L., Pisano, J. J. and Appela, E. (1973), *Biochem. Biophys. Res. Commun.*, **55**, 1220.
114. Graffeo, A. P., Haag, A. and Karger, B. L. (1973), *Anal. Lett.*, **6**, 505.
115. Frank, G. and Strubert, W. (1973), *Chromatographia*, **6**, 522.
116. Frankhauser, P., Fries, P., Stahala, P. and Brenner, M. (1974), *Helv. Chim. Acta*, **57**, 271.
117. De Vries, J. X., Frank, R. and Birr, Chr. (1975), *Febs. Lett.*, **55**, 65.
118. Matthews, E. W., Byfield, P. G. H. and MacIntyre, I. (1975), *J. Chromatogr.*, **110**, 369.
119. Bollet, C. and Caude, M. (1976), *J. Chromatogr.*, **121**, 323.
120. Haag, A. and Langer, K. (1974), *Chromatographia*, **7**, 659.
121. Ikekawa, N. and Koizumi, N. (1976), *J. Chromatogr.*, **119**, 227.
122. Casola, L., Di Matteo, G., Di Prisco, G. and Cervone, F. (1974), *Anal. Biochem.*, **57**, 38.
123. Seiler, N., Schneider, H. and Sonnenberg, K. D. (1970), *Fresenius'Z. Anal. Chem.*, **252**, 127.
124. Voelter, W. and Zech, K. (1975), *J. Chromatogr.*, **112**, 643.
125. Stahl, K. W., Schaefer, G. and Lamprecht, W. (1972), *J. Chromatogr. Sci.*, **10**, 95.
126. Scott, C. D., Chilcote, D. D., Katz, S. and Pitt Jr., W. W. (1973), *J. Chromatogr. Sci.*, **11**, 96.
127. Scott, C. D., Jolley, R. L., Pitt Jr., W. W. and Johnson, W. F. (1970), *Amer. J. Chem. Path.*, **53**, 701.
128. Majors, R. E. (1975), *Int. Lab.*, Nov/Dec. 11.
129. Rabel, F. M. (1973), *Anal. Chem.*, **45**, 957.
130. Sebestian, I. and Halasz, I. (1974), *Chromatographia*, **7**, 371.
131. Locke, D. C., Schmermund, J. T. and Banner, B. (1972), *Anal. Chem.*, **44**, 90.
132. Unger, K., Schier, G. and Beisel, V. (1973), *Chromatographia*, **6**, 456.
133. Weigand, N., Sebestian, I. and Halasz, I. (1974), *J. Chromatogr.*, **102**, 325.

134. Saunders, D. H., Barford, R. A., Magidman, P., Olszewski, L. T. and Rothbart, H. L. (1974), *Anal. Chem.*, **46**, 834.
135. Barford, R. A., Olszewski, L. T., Saunders, D. H., Magidman, P. and Rothbart, H. L. (1974), *J. Chromatogr. Sci.*, **12**, 555.
136. Kirkland, J. J. (1971), *J. Chromatogr. Sci.*, **9**, 206.
137. Kirkland, J. J. and Yates, P. C., (1973), U.S. Patent 3,722,181.
138. Pryde, A. (1974), *J. Chromatogr. Sci.*, **12**, 486.
139. Rehak, V. and Smolkova, E. (1976), *Chromatographia*, **9**, 219.
140. Locke, D. C. (1973), *J. Chromatogr. Sci.*, **11**, 120.
141. Leitch, R. E. and DeStefano, J. J. (1973), *J. Chromatogr. Sci.*, **11**, 105.
142. Rabel, F. M. (1975), *Int. Lab.*, July/Aug. 35.
143. Grushka, E. (ed.) (1974), *Bonded Stationary Phases in Chromatography*, Ann Arbor, Michigan.
144. Majors, R. E. and Hopper, M. J. (1974), *J. Chromatogr. Sci.*, **12**, 767.
145. Kirkland, J. J. (1975), *Chromatographia*, **8**, 661.
146. Gilpin, R. K., Korpi, J. A. and Janicki, C. A. (1975), *Anal. Chem.*, **47**, 1498.
147. Grushka, E. and Kikta Jr., E. J. (1974), *Anal. Chem.*, **46**, 1370.
148. Gilpin, R. K., Korpi, J. A. and Janicki, C. A. (1974), *Anal. Chem.*, **46**, 1314.
149. Schwarzenbach, R. (1976), *J. Chromatogr.*, **117**, 206.
150. Knox. J. H. and Pryde, A. (1975), *J. Chromatogr.*, **112**, 171.
151. Karch, K., Sebastian, I. and Halasz, I. (1976), *J. Chromatogr.*, **122**, 3.
152. Karch, K., Sebastian, I., Halasz, I. and Engelhardt, H. (1976), *J. Chromatogr.*, **122**, 171.
153. Vivilecchia, R. V., Cotter, R. L., Limpert, R. J., Thimot, N. Z. and Little, J. N. (1974), *J. Chromatogr.*, **99**, 407.
154. Unger, K., Becker, N. and Roumeliotis, P., (1976), *J. Chromatogr.*, **125**, 115.
155. Bakalyar, S. R., McIlwrick, R. and Roggendorf, E. (1977), *J. Chromatogr.*, **142**, 353.
156. Asmus, P. A., Low, C.-E. and Novotny, M. (1976), *J. Chromatogr.*, **119**, 25.
157. Chang, S. H., Gooding, K. M. and Regnier, F. E. (1976), *J. Chromatogr.*, **120**, 321.
158. Kingston, D. G. I. and Gerhart, B. B. (1976), *J. Chromatogr.*, **116**, 182.
159. Hunt, D. C., Wild, P. J. and Crosby, N. T. (1977), *J. Chromatogr.*, **130**, 320.
160. Knox, J. H. and Vasvari, G. (1973), *J. Chromatogr.*, **83**, 181.
161. Done, J. N., Knox, J. H. and Loheac, J. *Applications of High-Speed Liquid Chromatography*, Wiley-Interscience New York, (1974).
162. Kirkland, J. J., (1972), *J. Chromatogr. Sci.*, **10**, 129.
163. Majors, R. E., (1972), *Anal. Chem.*, **44**, 1722.
164. Cassidy, R. M., LeGay, D. S. and Frei, R. W. (1974), *Anal. Chem.*, **46**, 340.
165. Strubert, W. (1973), *Chromatographia*, **6**, 50.
166. Asshauer, J. and Halasz, I. (1974), *J. Chromatogr. Sci.*, **12**, 139.
167. Kirkland, J. J. (1972), *J. Chromatogr. Sci.*, **10**, 593.
168. Bristow, P. A., Brittain, P. N., Riley, C. M. and Williamson, B. F. (1977), *J. Chromatogr.*, **131**, 57.
169. Karger, B. L., Martin, M. and Guiochon, G. (1974), *Anal. Chem.*, **46**, 1640.
170. Little, J. N. and Fallick, G. J. (1975), *J. Chromatogr.*, **112**, 389.

171. Kirkland, J. J. (1973), *J. Chromatogr.*, **83**, 149.
172. Scott, R. P. W. and Kucera, P. (1974), *J. Chromatogr. Sci.*, **12**, 473.
173. Majors, R. E. (1972), *Anal. Chem.*, **44**, 1722.
174. Done, J. N. (1976), *J. Chromatogr.*, **125**, 43.
175. Saunders, D. L. (1976), *J. Chromatogr.*, **125**, 163.
176. Scott, R. P. W. and Kucera, P. (1976), *J. Chromatogr.*, **119**, 467.
177. Wolf III, J. P. (1973), *Anal. Chem.*, **45**, 1248.
178. Godbille, E. and Devaux, P. (1976), *J. Chromatogr.*, **122**, 317.
179. Godbille, E. and Devaux, P. (1974), *J. Chromatogr. Sci.*, **12**, 564.
180. Sie, S. T. and Van den Hoed, N. (1969), *J. Chromatogr. Sci.*, **7**, 257.
181. Baker, D. R., Henry, R. A., Williams, R. C., Hudson, D. R. and Parris, N. A. (1973), *J. Chromatogr.*, **83**, 233.
182. Bristow, P. A. (1976), *J. Chromatogr.*, **122**, 277.
183. Larmann, J. P., Williams, R. C. and Baker, D. R. (1975), *Chromatographia*, **8**, 92.
184. Little, J. N., Cotter, R. L., Prendergast, J. A. and McDonald, P. D. (1976), *J. Chromatogr.*, **126**, 439.
185. Conroe, K. E. (1975), *Chromatographia*, **8**, 119.
186. Attebery, J. A. (1975), *Chromatographia*, **8**, 121.
187. Eksborg, S. and Schill, G. (1973), *Anal. Chem.*, **45**, 2092.
188. Boehm, H.-P. (1966), *Angew. Chem. Int. Ed. Engl.*, **5**, 533.
189. Iler, R. K. (1955), *The Colloid Chemistry of Silica and the Silicates*, Cornell University Press, Ithaca, New York.
190. Unger, K. (1972), *Angew. Chem. Int. Ed. Engl.*, **11**, 267.
191. Scott, R. P. W. and Kucera, P. (1975), *J. Chromatogr. Sci.*, **13**, 337.
192. Scott, R. P. W. and Kucera, P. (1975), *J. Chromatogr.*, **112**, 425.
193. Soczewinski, E. (1969), *Anal. Chem.*, **41**, 179.
194. Snyder, L. R. (1974), *Anal. Chem.*, **46**, 1384.
195. Scott, R. P. W. (1976), *Contemporary Liquid Chromatography*, Wiley-Interscience, New York.
196. Guillemin, C. L., Thomas J. P., Thiault, S. and Bounine, J. P. (1977), *J. Chromatogr.*, **142**, 321.
197. Vermont, J., Deleuil, M., De Vries, A. J. and Guillemin, C. L. (1975), *Anal. Chem.*, **47**, 1329.
198. Engelhardt, H. and Wiedemann, H. (1973), *Anal. Chem.*, **45**, 1641.
199. Saunders, D. L. (1974), *Anal. Chem.*, **46**, 470.
200. El Rassi, Z., Gonnet, C. and Rocca, J. L. (1976), *J. Chromatogr.*, **125**, 179.
201. Scott, R. P. W. and Kucera, P. (1973), *Anal. Chem.*, **45**, 749.
202. Kennedy, G. J. and Knox, J. H. (1972), *J. Chromatogr. Sci.*, **10**, 549.
203. Kirkland, J. J. (1969), *J. Chromatogr. Sci.*, **7**, 361.
204. Kirkland, J. J. and Dilks Jr., C. H. (1973), *Anal. Chem.*, **45**, 1778.
205. Parris, N. A. (1974), *J. Chromatogr. Sci.*, **12**, 753.
206. Engelhardt, H., Asshauer, J., Neue, U. and Weigand, N. (1974), *Anal. Chem.*, **46**, 336.
207. Howard, G. A. and Martin, A. J. P. (1950), *Biochem. J.*, **56**, 532.
208. Telepchak, M. J. (1973), *Chromatographia*, **6**, 234.
209. Schmit, J. A., Henry, R. A., Williams, R. C. and Dieckman, J. F. (1971), *J. Chromatogr. Sci.*, **9**, 645.
210. Locke, D. C. (1974), *J. Chromatogr. Sci.*, **12**, 433.
211. Kikta Jr., E. J. and Grushka, E. (1976), *Anal. Chem.*, **48**, 1098.
212. Horgan, D. F. and Little, J. N. (1972), *J. Chromatogr. Sci.*, **10**, 76.

213. Karger, B. L., Gant, J. R., Hartkopf, A. and Weiner, P. H. (1976), *J. Chromatogr.*, **128**, 65.
214. Colin, H. and Guiochon, G. (1976), *J. Chromatogr.*, **126**, 43.
215. Cohn, W. E. (1949), *Science*, **109**, 377.
216. Moore, S. and Stein, W. H. (1951), *J. Biol. Chem.*, **192**, 663.
217. Khym, J. X. and Zill, L. P. (1951), *J. Amer. Chem. Soc.* **73**, 2399.
218. Helfferich, F. (1962), *Ion Exchange*, McGraw-Hill, New York.
219. Samuelson, O. (1963), *Ion Exchange Separations in Analytical Chemistry*, Wiley-Interscience, New York.
220. Scott, C. D. (1971), in *Modern Practice of Liquid Chromatography*, (ed. J. J. Kirkland), Wiley-Interscience, New York, Chapter 8.
221. Parris, N. A. (1976), *Instrumental Liquid Chromatography*, Elsevier, Amsterdam, Chapter 9.
222. Ertingshausen, G., Adler, H. J. and Reichler, A. S. (1969), *J. Chromatogr.*, **42**, 355.
223. Fransson, B., Wahlund, K.-G., Johansson, I. M. and Schill, G. (1976), *J. Chromatogr.*, **125**, 327.
224. Knox, J. H. and Laird, G. R. (1976), *J. Chromatogr.*, **122**, 17.
225a. *Bibliography on Liquid Exclusion Chromatography* (Gel Permeation Chromatography) 1972–1975. Sponsored by section D20.70.04 on Gel Permeation Chromatography of Subcommittee D.20.70. on Analytical Methods. Atomic and Molecular Data Series AMD 40–51. American Society for Testing Materials, Philadelphia, 1975.
225b. Gaylor, V. F., James, H. L. and Weetall, H. H. (1976), *Anal. Chem.*, **48**, 44R.
226. Parris, N. A. (1976), *Instrumental Liquid Chromatography*, Elsevier, Amsterdam, Chapter 10.
227. Bombaugh, K. J. (1971), in *Modern Practice of Liquid Chromatography*, (ed. J. J. Kirkland), Wiley-Interscience, New York, Chapter 7
228. Anderson, D. M. W. (1976), in *Practical High Performance Liquid Chromatography*, (ed. C. F. Simpson), Heyden, London, p. 153.
229. Billingham, N. C. (1976), in *Practical High Performance Liquid Chromatography*, (ed. C. F. Simpson), Heyden, London, p. 167.
230. Unger, K., Kern, R., Ninou, M. C. and Krebs, K.-F. (1974), *J. Chromatogr.*, **99**, 435.
231. Kirkland, J. J. (1976), *J. Chromatogr.*, **125**, 231.
232. Reiner, R. H. and Walch, A. (1971), *Chromatographia*, **4**, 578.
233. Guilford, H. (1973), *Chem. Soc. Rev.*, **2**, 249.
234. Ripphann, J. and Halpaap, H. (1975), *J. Chromatogr.* **112**, 81.
235. Bailey, F. and Brittain, P. N. (1973), *J. Chromatogr.*, **83**, 431.
236. Olson, M. C. (1973), *J. Pharm. Sci.*, **62**, 2001.
237. Henry, R. A. and Schmit, J. A. (1970), *Chromatographia*, **3**, 116.
238. Brown, N. D., Lofberg, R. T. and Gibson, T. P. (1974), *J. Chromatogr.* **99**, 635.
239. Weinberger, N. and Chidsey, C. (1975), *Clin. Chem.* **21**, 834.
240. Chen, C-T, Hayakawa, K., Imanari, T. and Tamura, Z. (1975), *Chem. Pharm. Bull.*, **23**, 2173.
241. Buhs, R. P., Maxim, T. E., Allen, N., Jacob, T. A. and Wolf, F. J. (1974), *J. Chromatogr.*, **99**, 609.
242. Tsuji, K. and Robertson, J. H. (1975), *J. Pharm. Sci.*, **64**, 1542.
243. Wragg, J. S. and Johnson, G. W. (1974), *Pharm. J.*, **213**, 601 and 613.
244. Bailey, F. (1976), *J. Chromatogr.*, **12**, 73.

245. Dixon, P. F., Stoll, M. S. and Lim, C. K. (1976), *Ann. Clin. Biochem.*, **13**, 409.
246. Wheals, B. B. (1976), *J. Chromatogr.*, **122**, 85.
247. Blaha, J. M., Knevel, A. M. and Hem, S. L. (1975), *J. Pharm. Sci.*, **64**, 1384.
248. Weber, D. J. (1972), *J. Pharm. Sci.*, **61**, 1797.
249. Riggin, R. M., Schmidt, A. L. and Kissinger, P. T. (1975), *J. Pharm. Sci.*, **64**, 680.
250. Korpi, J., Wittmer, D. P., Sandmann, B. J. and Haney, W. G. (1976), *J. Pharm. Sci.*, **65**, 1087.
251. Bracey, A. (1973), *Drug Standards*, **62**, 1695.
252. Anders, M. W. and Latorre, J. P. (1970), *Anal. Chem.*, **42**, 1430.
253. Tjaden, U. R., Kraak, J. C. and Huber, J. F. K. (1977), *J. Chromatogr., (Biomed. Appl.)*, **143**, 183.
254. Vigh, G. and Inczedy, J. (1974), *J. Chromatogr.*, **102**, 381.
255. Vigh, G. and Inczedy, J. (1976), *J. Chromatogr.*, **116**, 472.
256. White, E. R., Carroll, M. A., Zarembo, J. E. and Bender, A. D. (1975), *J. Antibiotics*, **28**, 205.
257. Tsuji, K. and Robertson, J. H. (1975), *J. Chromatogr.* **112**, 663.
258. Knox, J. H. and Jurand, J. (1975), *J. Chromatogr.*, **110**, 103.
259. Chevalier, G., Bollet, C., Rohrbach, P., Risse, C., Caude, M. and Rosset, R. (1976), *J. Chromatogr.*, **124**, 343.
260. Tsuji, K. and Robertson, J. H. (1976), *J. Pharm. Sci.*, **65**, 400.
261. Onishi, H. R., Daoust, D. R., Zimmerman, S. B., Hendlin, D. and Stapley, E. O. (1974), *Antimicrob. Ag. Chemother.*, **5**, 38.
262. Cooper, M. J., Anders, M. W. and Mirkin, B. L., (1973), *Drug Metab. Dispos.*, **1**, 569.
263. Tsuji, K., Robertson, J. H. and Bach, J. A. (1974), *J. Chromatogr.*, **99**, 597.
264. Rzeszotarski, W. J. and Mauger, A. B. (1973), *J. Chromatogr.*, **86**, 246.
265. Mechlinski, W. and Schaffner, C. P. (1974), *J. Chromatogr.*, **99**, 619.
266. Hansen, S. H. and Thomsen, M. (1976), *J. Chromatogr.* **123**, 205.
267. Omura, S., Suzuki, Y., Nakagawa, A. and Hata, T. (1973), *J. Antibiotics*, **26**, 794.
268. Tsuji, K., Robertson, J. H. and Beyer, W. F. (1974), *Anal. Chem.*, **46**, 539.
269. Ascione, P. P., Zagar, J. B. and Chrekian, G. P., (1972), *J. Chromatogr.*, **65**, 377.
270. Butterfield, A. G., Hughes, D. W., Wilson, W. L. and Pound, N. J. (1975), *J. Pharm. Sci.*, **64**, 316.
271. Lindauer, R. F., Cohen, D. M., and Munnelly, K. P. (1976), *Anal. Chem.*, **48**, 1731.
272. Magic, S. E. (1976), *J. Chromatogr.*, **129**, 73.
273. Tsuji, K. and Robertson, J. H. (1974), *J. Chromatogr.*, **94**, 245.
274. Mays, D. L., Van Apeldoorn, R. J. and Lauback, R. G., (1976), *J. Chromatogr.*, **120**, 93.
275. Hulhoven, R. and Desager, J. P. (1976), *J. Chromatogr.* **125**, 369.
276. Vigh, G. and Inczedy, J. (1976), *J. Chromatogr.*, **129**, 81.
277. Papp, E., Magyar, K. and Schwarz, H. J. (1976), *J. Pharm. Sci.*, **65**, 441.
278. *The United States Pharmacopeia*, 18th Rev., Mack Publishing Co., Easton Pa., (1970). p. 760.
279. Kram, T. C. (1972), *J. Pharm. Sci.*, **61**, 254.

280. Poet, R. B. and Pu, H. H. (1973), *J. Pharm. Sci.*, **62**, 809.
281. Karger, B. L., Su, S. C., Marchese, S. and Persson, B. A. (1974), *J. Chromatogr. Sci.*, **12**, 678.
282. Su, S. C., Hartkopf, A. V. and Karger, B. L. (1976), *J. Chromatogr.*, **119**, 523.
283. Rader, B. R. (1973), *J. Pharm. Sci.*, **62**, 1148.
284. Cobb, P. H. and Hill, G. T. (1976), *J. Chromatogr.*, **123**, 444.
285. Bighley, L. D. and McDonnell, J. P. (1975), *J. Pharm. Sci.*, **64**, 1549.
286. Penner, M. H. (1975), *J. Pharm. Sci.*, **64**, 1017.
287. Sharma, J. P., Perkins, E. G. and Bevill, R. F. (1976), *J. Pharm. Sci.*, **65**, 1606.
288. Cooper, M. J., Anders, M. W., Sinaiko, A. R. and Mirkin, B. L. (1976), in *High Pressure Liquid Chromatography In Clinical Chemistry* (ed. P. F. Dixon, C. H. Gray, C. K. Lim and M. S. Stoll), Academic Press, London, New York, San Francisco, p. 175.
289. Johnson, K. L., Jeter, D. T. and Claiborne, R. C., (1975), *J. Pharm. Sci.*, **64**, 1657.
290. Bailey, F., Brittain, P. N. and Williamson, B. F. (1975), *J. Chromatogr.*, **109**, 305.
291. Sondack, D. L. and Koch, W. L. (1977), *J. Chromatogr.*, **132**, 352.
292. Stewart, J. T., Honigberg, I. L., Brant, J. P., Murray, W. A., Webb, J. L. and Smith, J. B. (1976), *J. Pharm. Sci.*, **65**, 1536.
293. Blair, A. D., Forrey, A. W., Meijsen, B. T. and Cutler, R. E. (1975), *J. Pharm. Sci.*, **64**, 1334.
294. Shargel, L., Koss, R. F., Crain, A. V. R. and Boyle, V. J. (1973), *J. Pharm. Sci.*, **62**, 1453.
295. Knox, J. H. and Jurand, J. (1975), *J. Chromatogr.*, **103**, 311.
296. Detaevernier, M. R., Dryon, L. and Massart, D. L. (1976), *J. Chromatogr.*, **128**, 204.
297. Watson, I. D. and Stewart, M. J. (1975), *J. Chromatogr.* **110**, 389.
298. Gonnet, C. and Rocca, J. L. (1976), *J. Chromatogr.*, **120**, 419.
299. Mellstrom, B. and Eksborg, S. (1976), *J. Chromatogr.*, **116**, 475.
300. Wu, C.-Y, and Siggia, S (1972), *Anal. Chem.*, **44**, 1499.
301. Nelson, J. J. (1973), *J. Chromatogr. Sci.*, **11**, 28.
302. Manion, C. V., Shoeman, D. W. and Azarnoff, D. L. (1974), *J. Chromatogr.*, **101**, 169.
303. Franconi, L. C., Hawk, G. L., Sandmann, B. J. and Haney, W. G. (1976), *Anal. Chem.*, **48**, 372.
304. Situr, D. S., Piafsky, K. M., Rangno, R. E. and Ogilvie, R. I. (1975), *Clin. Chem.*, **21**, 1774.
305. Weddle, O. H. and Mason, W. D. (1976), *J. Pharm. Sci.*, **65**, 865.
306. Murgia, E., Richards, P. and Walton, H. F., (1973), *J. Chromatogr.*, **87**, 523.
307. Wildanger, W., (1975), *J. Chromatogr.*, **114**, 480.
308. Pachla, L. A. and Kissinger, P. T. (1975), *Clin. Chim. Acta*, **59**, 309.
309. Slaunwhite, W. D., Pachla, L. A., Wenke, D. C. and Kissinger, P. T. (1975), *Clin. Chem.*, **21**, 1427.
310. Evans, J. E. (1973), *Anal. Chem.*, **45**, 2428.
311. Roos, R. W. (1972), *J. Pharm. Sci.*, **61**, 1979.
312. Atwell, S. H., Green, V. A. and Haney, W. G. (1975), *J. Pharm. Sci.*, **64**, 806.
313. Needham, L. L. and Kochhar, M. M. (1975), *Clin. Chem.*, **21**, 169.

314. Needham, L. L. and Kochhar, M. M. (1975), *J. Chromatogr*, **111**, 422.
315. Eichelbaum, M. and Bertilsson, L. (1975), *J. Chromatogr.*, **103**, 135.
316. Westenberg, H. G. M. and De Zeeuw, R. A. (1976), *J. Chromatogr.*, **118**, 217.
317. Adams, R. F. and Vandemark, F. L. (1976), *Clin. Chem.*, **22**, 25.
318. Macek, K. and Rehak, V. (1975), *J. Chromatogr.*, **105**, 182.
319. Scott, C. G. and Bommer, P. (1970), *J. Chromatogr. Sci.*, **8**, 446.
320. Rodgers, D. H. (1974), *J. Chromatogr. Sci.*, **12**, 742.
321. Huettemann, R. E. and Shroff, A. P. (1975), *J. Pharm. Sci.*, **64**, 1339.
322. Twitchett, P. J., Gorvin, A. E. P. and Moffat, A. C. (1976), *J. Chromatogr.*, **120**, 359.
323. Bugge, A. (1976), *J. Chromatogr.*, **128**, 111.
324. Harzer, K. and Barchet, R. (1977), *J. Chromatogr.*, **132**, 83.
325. Lindner, W., Frei, R. W. and Santi, W. (1975), *J. Chromatogr.*, **108**, 299.
326. Caude, M., Lefevre, J. P. and Rosset, R. (1975), *Chromatographia*, **8**, 217.
327. Byrne, M. J. (1976), *J. Assoc. Offic. Anal. Chem.*, **59**, 693.
328. Lindner, W., Frei, R. W. and Santi, W. (1975), *J. Chromatogr.*, **111**, 365.
329. Stevenson, R. L. and Burtis, C. A. (1971), *J. Chromatogr.*, **61**, 253.
330. Larson, P., Murgia, E., Hsu, T.-J. and Walton, H. F., (1973), *Anal. Chem.*, **45**, 2306.
331. Cox, G. S., Loscombe, C. R., Slucutt, M. J., Sugden, K. and Upfield, J. A. (1976), *J. Chromatogr.*, **117**, 269.
332. Ascione, P. P. and Chrekian, G. P. (1975), *J. Pharm. Sci.*, **64**, 1029.
333. Stillman, R. and Ma, T. S. (1974), *Mikrochimica Acta*, 641.
334. Rosenbaum, D. (1974), *Anal. Chem.*, **48**, 2226.
335. Caude, M. and LePhan, X. (1976), *Chromatographia*, **9**, 20.
336. Gilpin, R. K., Korpi, J. A. and Janicki, C. A. (1975), *J. Chromatogr.*, **107**, 115.
337. Honigberg, I. L., Stewart, J. T. and Smith, A. P. (1974), *J. Pharm. Sci.*, **63**, 766.
338. Menyharth, A., Mahn, F. P. and Heveran, J. E. (1974), *J. Pharm. Sci.*, **63**, 431.
339. Sprieck, T. L. (1974), *J. Pharm. Sci.*, **63**, 591.
340. Higgins, J. W. (1975), *J. Chromatogr.*, **115**, 232.
341. Tscherne, R. J. and Capitano, G. (1977), *J. Chromatogr.*, **136**, 337.
342. Pound, N. J., McGilveray, I. J. and Sears, R. W. (1974), *J. Chromatogr.*, **89**, 23.
343. Pound, N. J. and Sears, R. W. (1975), *J. Pharm. Sci.*, **64**, 284.
344. Mrochek, J. E., Katz, S., Christie, W. H. and Dinsmore, S. R. (1974), *Clin. Chem.*, **20**, 1086.
345. Howie, D. Adriaenssens, P. H. and Prescott, L. F. (1977), *J. Pharm. Pharmacol.*, **29**, 235.
346. Wong, L. T., Solomonraj, G. and Thomas, B. H. (1976), *J. Pharm. Sci.*, **65**, 1064.
347. Knox, J. H. and Jurand, J. (1977), *J. Chromatogr.*, **142**, 651.
348. Duggin, G. G. (1976), *J. Chromatogr.*, **121**, 156.
349. Skellern, G. G. and Salole, E. G. (1975), *J. Chromatogr.*, **114**, 483.
350. Roggia, S., Grossoni, G., Pelizza, G., Ratti, B. and Gallo, G. G. (1976), *J. Chromatogr.*, **124**, 169.
351. Honigberg, I. L., Stewart, J. T., Smith, A. P. and Hester, D. W. (1975), *J. Pharm. Sci.*, **64**, 1201.

352. Honigberg, I. L., Stewart, J. T., Smith, A. P., Plunkett, R. D. and Hester, D. W. (1974), *J. Pharm. Sci.*, **63**, 1762.
353. Moskalyk, R. E., Locock, R. A., Chatten, L. G., Veltman, A. M. and Bielech, M. F. (1975), *J. Pharm. Sci.*, **64**, 1406.
354. Cohen, D. M. and Munnelly, K. P. (1976), *J. Pharm. Sci.*, **65**, 1413.
355. Williamson, D. E. (1976), *J. Pharm. Sci.*, **65**, 138.
356. MacDougall, M. L., Shoeman, D. W. and Azarnoff, D. L., (1975), *Res. Comm. Chem. Pathol. Pharmacol.*, **10**, 285.
357. Chu, S-Y, and Sennello, L. T. (1976), *J. Chromatogr.*, **117**, 415.
358. Christophersen, A-S., Rasmussen, K. E. and Salvesen, B. (1977), *J. Chromatogr.*, **132**, 91.
359. Bayne, W. F., Rogers, G. and Crisologo, N. (1975), *J. Pharm. Sci.*, **64**, 402.
360. Cashman, P. J. Thornton, J. I. and Shelman, D. L., (1973), *J. Chromatogr. Sci.*, **11**, 7.
361. Jane, I. (1975), *J. Chromatogr.*, **111**, 227.
362. Needham, L. L. and Kochhar, M. M. (1975), *J. Chromatogr.*, **114**, 220.
363. Wu, C-Y, Siggia, S., Robinson, T. and Waskiewicz, R. D. (1973), *Anal. Chim. Acta.*, **63**, 393.
364. Verpoorte, R. and Svendsen, A. B. (1974), *J. Chromatogr.*, **100**, 227.
365. Knox, J. H. and Jurand, J. (1973), *J. Chromatogr.*, **82**, 398.
366. Knox, J. H. and Jurand, J. (1973), *J. Chromatogr.*, **87**, 95.
367. Jane, I. and Taylor, J. F. (1975), *J. Chromatogr.*, **109**, 37.
368. Twitchett, P. J. (1975), *J. Chromatogr.*, **104**, 205.
369. Wheals, B. B. and Smith, R. N. (1975), *J. Chromatogr.*, **105**, 396.
370. Schmit, J. A., Henry, R. A., Williams, R. C. and Dieckman, J. F. (1971), *J. Chromatogr. Sci.*, **9**, 645.
371. Smith, R. N. (1975), *J. Chromatogr.*, **115**, 101.
372. Abbott, S. R., Abu-Shumays, A., Loeffler, K. O. and Forrest, I. S. (1975), *Chem. Pathol. Pharmacol.*, **10**, 9.
373. Jane, I. and Wheals, B. B. (1973), *J. Chromatogr.*, **84**, 181.
374. Wittwer, J. D. and Kluckhohn, J. H. (1973), *J. Chromatogr. Sci.*, **11**, 1.
375. Heacock, R. A., Langille, K. R., MacNeil, J. D. and Frei, R. W. (1973), *J. Chromatogr.*, **77**, 425.
376. Baker, D. R., Williams, R. C. and Steichen, J. C. (1974), *J. Chromatogr. Sci.*, **12**, 499.
377. Christie, J., White, M. W. and Wiles, J. M. (1976), *J. Chromatogr.*, **120**, 496.
378. Chan, M. L., Whetsell, C. and McChesney, J. D. (1974), *J. Chromatogr. Sci.*, **12**, 512.
379. Smith, D. W., Beasley, T. H., Charles, R. L. and Ziegler, H. W. (1973), *J. Pharm. Sci.*, **62**, 1691.
380. Stutz, M. H. and Sass, S. (1973), *Anal. Chem.*, **45**, 2134.
381. Verpoorte, R. and Svendsen, A. B. (1976), *J. Chromatogr.*, **120**, 203.
382. Honigberg, I. L., Stewart, J. T., Smith, A. P., Plunkett, R. D. and Justice, E. L. (1975), *J. Pharm. Sci.*, **64**, 1389.
383. Santi, W., Huen, J. M. and Frei, R. W. (1975), *J. Chromatogr.*, **115**, 423.
384. Kingston, D. G. I. and Li, B. T. (1975), *J. Chromatogr.*, **104**, 431.
385. Jolliffe, G. H. and Shellard, E. J. (1973), *J. Chromatogr.*, **81**, 150.
386. Perchalski, R. J., Winefordner, J. D. and Wilder, B. J. (1975), *Anal. Chem.*, **47** 1993.
387. Bethke, H., Delz, B. and Stich, K. (1976), *J. Chromatogr.*, **123**, 193.

388. Verpoorte, R. and Svendsen, A. B. (1974), *J. Chromatogr.*, **100**, 231.
389. Verpoorte, R. and Svendsen, A. B. (1975), *J. Chromatogr*., **109**, 441.
390. Frei, R. W., Santi, W. and Thomas, M. (1976), *J. Chromatogr.*, **116**, 365.
391. Beyer, W. F. (1972), *Anal. Chem.*, **44**, 1312.
392. Molins, D., Wong, C. K., Cohen, D. M. and Munnelly, K. P. (1975), *J. Pharm. Sci.*, **64**, 123.
393. Sved, S., McGilveray, I. J. and Beaudoin, N. (1976), *J. Pharm. Sci.*, **65**, 1356.
394. Weber, D. J. (1976), *J. Pharm. Sci.*, **65**, 1502.
395. Mollica, J. A., Padmanabhan, G. R. and Strusz, R. (1973), *Anal. Chem.*, **45**, 1859.
396. Inaba, T., Besley, M. E. and Chow, E. J. (1975), *J. Chromatogr.*, **104**, 165.
397. Lecaillon, J-B. and Souppart, C. (1976), *J. Chromatogr.*, **121**, 227.
398. Hoye, A. (1967), *J. Chromatogr.*, **28**, 379.
399. Falk, A. J. (1974), *J. Pharm. Sci.*, **63**, 274.
400. Bighley, L. D., Wurster, D.,E., Cruden-Loeb, D. and Smith, R. V. (1975), *J. Chromatogr.*, **110**, 375.
401. Meffin, P. J., Harapat, S. R. and Harrison, D. C., (1977), *J. Chromatogr.*, **132**, 503.
402. Carr, K., Woosley, R. L. and Oates, J. A. (1976), *J. Chromatogr.*, **129**, 363.
403. Khalil, S. K. W. and Shelver, W. H. (1976), *J. Pharm. Sci.*, **65**, 606.
404. Urbanyi, T., Piedmont, A., Willis, E. and Manning, G. (1976), *J. Pharm. Sci.*, **65**, 257.
405. Aitzetmüller, K. (1975), *J. Chromatogr. Sci.*, **13**, 454.
406. Takata, Y. and Muto, G. (1973), *Anal. Chem.*, **45**, 1864.
407. Politzer, I. R., Griffin, G. W., Dowty, B. J. and Laseter, J. L. (1973), *Anal. Lett.*, **6**, 539.
408. Evans, J. E. and McCluer, R. H. (1972), *Biochim. Biophys. Acta.*, **270**, 565.
409. Jungalwala, F. B., Turel, R. J., Evans, J. E. and McCluer, R. H. (1975), *J. Biochem.*, **145**, 517.
410. Cooper, M. J. and Anders, M. W. (1975), *J. Chromatogr. Sci.*, **13**, 407.
411. Aitzetmüller, K. (1975), *J. Chromatogr.*, **113**, 231.
412. Rouser, G., Kritchevski G. and Yamamoto, A. (1967), *Lipid Chromatographic Analysis*, Marcel Dekker Inc., New York, p. 99.
413. Kiuchi, K., Ohta, T. and Ebine, H. (1975), *J. Chromatogr. Sci.*, **13**, 461.
414. Stolyhwo, A. and Privett, O. S. (1973), *J. Chromatogr. Sci.*, (1973), **11**, 20.
415. Erdahl, W. L., Stolyhwo, A. and Privett, O. S. (1973), *J. Amer. Oil Chem. Soc.*, **50**, 513.
416. Privett, O. S., Dougherty, K. A., Erdahl, W. L. and Stolyhwo, A. (1973), *J. Amer. Oil Chem. Soc.*, **50**, 516.
417. Lawrence, J. G. (1973), *J. Chromatogr.*, **84**, 299.
418. Carter, T., Magnani, H. N. and Watts, R. (1976), in *High Pressure Liquid Chromatography in Clinical Chemistry*, (ed. P. F. Dixon, C. H. Gray, C. K. Lim and M. S. Stoll), Academic Press, London, New York, San Francisco, p. 33.
419. Fischer, R. (1974), *Chromatographia*, **7**, 207.
420. Sinsel, J. A., LaRue, B. M. and McGraw, L. D. (1975), *Anal. Chem.*, **47**, 1987.

421. Pei, P. T. S., Henly, R. S. and Ramachandran, S., (1975), *Lipids*, **10**, 152.
422. Mikes, F., Schurig, V. and Gil-Av., E., (1973), *J. Chromatogr.*, **83**, 91.
423. Schomburg, G. and Zegarski, K. (1975), *J. Chromatogr.*, **114**, 174.
424. Arvidson, G. A. E. (1975), *J. Chromatogr.*, **103**, 201.
425. Scholfield, C. R. (1975), *Anal. Chem.*, **47**, 1417.
426. Knapp, D. R. and Krueger, S. (1975), *Anal. Lett.*, **8**, 603.
427. Chan, H. W. S. and Prescott, F. A. A. (1975), *Biochim. Biophys. Acta.*, **380**, 141.
428. Fan, L. L., Masters, B. S. S. and Prough, R. A. (1976), *Anal. Biochem.*, **71**, 265.
429. McCluer, R. H., Evans, J. E. and Jungalwala, F. B. (1975), *Trans. Am. Soc. Neurochem.*, **6**, 183.
430. Jungalwala, F. B., Evans, J. E. and McCluer, R. H. (1976), *Biochem. J.*, **155**, 55.
431. Lairon, D., Amic, J., LaFont, H., Nalborne, G., Domingo, N. and Hauton, J. (1974), *J. Chromatogr.*, **88**, 183.
432. Sugita, M., Iwamori, M., Evans, J., McCluer, R. H., DuLaney, J. T. and Moser, H. W. (1974), *J. Lipid Res.*, **15**, 223.
433. Iwamori, M. and Moser, H. W. (1975), *Clin. Chem.*, **21**, 725.
434. McCluer, R. H. and Evans, J. E. (1973), *J. Lipid Res.*, **14**, 611.
435. O'Hare, M. J., Nice, E. C., Magee-Brown, R. and Bullman, H. (1976), *J. Chromatogr.*, **125**, 357.
436. Siggia, S. and Dishman, R. A., (1970), *Anal. Chem.*, **42**, 1223.
437. Huber, J. F. K., Hulsman, J. A. R. J. and Meijers, C. A. M., (1971), *J. Chromatogr.*, **62**, 79.
438. Lafosse, M., Keravis, G. and Durand, M. H. (1976), *J. Chromatogr.*, **118**, 283.
439. Huber, J. F. K., Alderlieste, E. T., Harren, H. and Poppe, H. (1973), *Anal. Chem.*, **45**, 1337.
440. Trefz, F. K., Byrd, D. J. and Kochen, W. (1975), *J. Chromatogr.*, **107**, 181.
441. Hesse, C. and Hövermann, W. (1973), *Chromatographia*, **6**, 345.
442. Hesse, C., Pletszik, K. and Hotzel, D. (1974), *Z. Klin. Chem. Klin. Biochem.*, **12**, 193.
443. Van den Berg, J. H. M., Milley, J., Vonk, N. and Deelder, R. S. (1977), *J. Chromatogr.*, **132**, 421.
444. Parris, N. A. (1974), *J. Chromatogr. Sci.*, **12**, 753.
445. Meijers, C. A. M., Hulsman, J. A. R. J. and Huber, J. F. K. (1972), *Z. Anal. Chem.*, **261**, 347.
446. Loo, J. C. K., Butterfield, A. G., Moffatt, J. and Jordan, N. (1977), *J. Chromatogr., (Biomed. Appl.)*, **143**, 275.
447. Touchstone, J. C. and Wortmann, W. (1973), *J. Chromatogr.*, **76**, 244.
448. Wortmann, W., Schnabel, C. and Touchstone, J. C. (1973), *J. Chromatogr.*, **84**, 396.
449. Krzeminski, L. F., Cox, B. L., Perrel, P. N. and Schiltz, R. A. (1972), *J. Agr. Food Chem.*, **20**, 970.
450. Loo, J. C. K. and Jordan, N. (1977), *J. Chromatogr., (Biomed. Appl.)*, **143**, 314.
451. Dolphin, R. J. (1973), *J. Chromatogr.*, **83**, 421.
452. Dolphin, R. J. and Pergande, P. J. (1977), *J. Chromatogr., (Biomed Appl.)*, **143**, 267.

453. Fantl, V., Lim, C. K. and Gray, C. H., (1976) in *High Pressure Liquid Chromatography in Clinical Chemistry*, (ed. P. F. Dixon, C. H. Gray, C. K. Lim and M. S. Stoll), Academic Press, London, New York, San Francisco, p. 51.
454. Butterfield, A. G., Lodge, B. A. and Pound, N. J. (1973), *J. Chromatogr. Sci.*, 11, 401.
455. Roos, R. W. (1976), *J. Chromatogr. Sci.*, 14, 505.
456. Roos, R. W. (1974), *J. Pharm. Sci.*, 63, 594.
457. Karger, B. L. and Berry, L. V. (1971), *Clin. Chem.*, 17, 757.
458. Butterfield, A. G., Lodge, B. A., Pound, N. J. and Sears, R. W. (1975), *Pharm. Anal.*, 64, 441.
459. Huetemann, R. E. and Shroff, A. P. (1975), *J. Chromatogr. Sci.*, 13, 357.
460. Slocum, S. A. and Studebaker, J. F. (1975), *Anal. Biochem.*, 68, 242.
461. Rees, H. H., Donnahey, P. L. and Goodwin, T. W. (1976), *J. Chromatogr.*, 116, 281.
462. Higgins, J. W. (1976), *J. Chromatogr.*, 121, 329.
463. Evans, F. J. (1974), *J. Chromatogr.*, 88, 411.
464. Castle, M. C. (1975), *J. Chromatogr.*, 115, 437.
465. Lindner, W. and Frei, R. W. (1976), *J. Chromatogr.* 117, 81.
466. Nachtmann, F., Spitzy, H. and Frei, R. W. (1976), *Anal. Chem.*, 48, 1576.
467. Nachtmann, F., Spitzy, H. and Frei, R. W. (1976), *J. Chromatogr.*, 122, 293.
468. Hunter, I. R., Walden, M. K., Wagner, J. R. and Heftmann, E. (1976), *J. Chromatogr.*, 119, 223.
469. Ryback, G. (1976), *J. Chromatogr.*, 116, 207.
470. Shaw, R. and Elliot, W. H. (1976), *Anal. Biochem.*, 74, 273.
471. Dunham, E. W. and Anders, M. W. (1973), *Prostaglandins*, 4, 85.
472. Morozowich, W. (1974), *J. Pharm. Sci.*, 63, 800.
473. Andersen, N. and Leovey, E. M. K. (1974), *Prostaglandins*, 6, 361.
474. Carr, K., Sweetman, B. J. and Frolich, J. C. (1976), *Prostaglandins*, 11, 3.
475. Morozowich, W. and Douglas, S. L. (1975), *Prostaglandins*, 10, 19.
476. Fitzpatrick, F. A. (1976), *Anal. Chem.*, 48, 499.
477. Lustgarten, R. K. (1976), *J. Pharm. Sci.*, 65, 1533.
478. Weinshenker, N. M. and Longwell, A. (1972), *Prostaglandins*, 2, 207.
479. Roseman, T. J., Butler, S. S. and Douglas, S. L. (1976), *J. Pharm. Sci.*, 65, 673.
480. Fitzpatrick, F. A. (1976), *J. Pharm. Sci.*, 65, 1609.
481. Funasaka, W., Hanai, T. and Fujimura, K. (1974), *J. Chromatogr. Sci.*, 12, 517.
482. Grushka, E., Durst, H. D. and Kikta Jr., E. J. (1975), *J. Chromatogr.*, 112, 673.
483. Richards, M. (1975), *J. Chromatogr.*, 115, 259.
484.– Katz, S. and Pitt, W. W. (1972), *Anal. Lett.*, 5, 177.
485. Katz, S., Pitt, W. W. and Jones, G. (1973), *Clin. Chem.*, 19, 817.
486. Carlsson, B. and Samuelson, O. (1970), *Anal. Chim. Acta*, 49, 247.
487. Hyakutake, H. and Hanai, T. (1975), *J. Chromatogr.*, 108, 385.
488. Hayashi, T., Sugiura, T., Terada, H., Kawai, S. and Ohno, T. (1976), *J. Chromatogr.*, 118, 403.
489. Kasai, Y., Ozawa, Y., Tanimura, T. and Tamura, Z. (1975), *Anal. Chem.*, 47, 34.
490. Nakajima, M., Ozawa, Y., Tanimura, T. and Tamura, Z. (1976), *J. Chromatogr.*, 123, 129.

491. Bishop, C. T. (1955), *Can. J. Chem.*, **33**, 1073.
492. Hough, L., Jones, J. K. N. and Wadman, W. H. (1949), *J. Chem. Soc.*, 2511.
493. Whistler, R. L. and Durso, D. F. (1950), *J. Amer. Chem. Soc.*, **72**, 677.
494. Whistler, R. L. and Tu, C. C. (1952), *J. Amer. Chem. Soc.*, **74**, 3609.
495. John, M., Trenel, G. and Dellweg, H. (1969), *J. Chromatogr.*, **42**, 476.
496. Chitumbo, K. and Brown, W. (1971), *J. Polym. Sci.*, **36**. 279.
497. Brown, W. and Andersson, O. (1972), *J. Chromatogr.*, **67**, 163.
498. Belue, G. P. and McGinnis, G. D. (1974), *J. Chromatogr.*, **97**, 25.
499. Jandera, P. and Churacek, J. (1974), *J. Chromatogr.*, **98**, 55.
500. Khym, J. X. and Zill, L. P. (1952), *J. Amer. Chem. Soc.*, **74**, 2090.
501. Kessler, R. B. (1967), *Anal. Chem.*, **39**, 1416.
502. Floridi, A. (1971), *J. Chromatogr.*, **59**, 61.
503. Hough, L., Jones, J. V. S. and Wusteman, P. (1972), *Carbohydr. Res.*, **21**, 9.
504. Voelter, W. and Bauer, H. (1972), *J. Chromatogr.*, **126**, 693.
505. Jones, J. K. N., Wall, R. A. and Pittet, A. O. (1960), *Can. J. Chem.*, **38**, 2285.
506. Jones, J. K. N. and Wall, R. A. (1960), *Can. J. Chem.*, **38**, 2290.
507. Martinsson, E. and Samuelson, O. (1970), *J. Chromatogr.*, **50**, 429.
508. Havlicek, J. and Samuelson, O. (1972), *Carbohydr. Res.*, **22**, 307.
509. Havlicek, J. and Samuelson, O. (1974), *Chromatographia*, **7**, 361.
510. Havlicek, J. and Samuelson, O. (1975), *Anal. Chem.*, **47**, 1854.
511. Goulding, R. W. (1975), *J. Chromatogr.*, **103**, 229.
512. Lawrence, J. F. (1975), *Chimia*, **29**, 367.
513. Walborg, E. F. and Christensson, L. (1965), *Anal. Biochem.*, **13**, 186.
514. Jonsson, P. and Samuelson, O. (1966), *Science Tools*, **13**, 17.
515. Davies, A. M. C., Robinson, D. S. and Couchman, R. (1974), *J. Chromatogr.*, **101**, 307.
516. Palmer, J. K. (1975), *Anal. Lett.*, **8**, 215.
517. Linden, J. C. and Lawhead, C. L. (1975), *J. Chromatogr.*, **105**, 125.
518. Rocca, J. L. and Rouchouse, A. (1976), *J. Chromatogr.*, **117**, 216.
519. Lehrfeld, J. (1976), *J. Chromatogr.*, **120**, 141.
520. Nachtmann, F. and Budna, K. W. (1977), *J. Chromatogr.*, **136**, 279.
521. Belue, G. P. (1974), *J. Chromatogr.*, **100**, 233.
522. Dark, W. A., Conrad, E. C. and Crossman, L. W. (1974), *J. Chromatogr.*, **91**, 247.
523. Ward, R. S. and Pelter, A. (1974), *J. Chromatogr. Sci.* **12**, 570.
524. Cegla, G. F. and Bell, K. R. (1977), *J. Amer. Oil Chem. Soc.*, **54**, 150.
525. Rabel, F. M., Caputo, A. G. und Butts, E. T. (1976), *J. Chromatogr.*, **126**, 731.
526. Jolley, R. L., Warren, K. S., Scott, C. D., Jainchill, J. L. and Freeman, M. L. (1970), *Amer. J. Clin. Path.*, **53**, 793.
527. Jolley, R. L. and Freeman, M. L. (1968), *Clin. Chem.*, **14**, 538.
528. Burtis, C. A. (1970), *J. Chromatogr.*, **52**, 97.
529. Meagher, R. B. and Furst, A. (1976), *J. Chromatogr.*, **117**, 211.
530. Wall. R. A. (1971), *J. Chromatogr.*, **60**, 195.
531. Mori, K. (1975), *Jap. J. Ind. Health*, **17**, 116.
532. Mori, K. (1975), *Jap. J. Ind. Health*, **17**, 170.
533. Kissinger, P. T., Riggin, R. M., Alcorn, R. L. and Rau, L-D. (1975), *Biochem. Med.*, **13**, 299.
534. Seki, T. (1976), *J. Chromatogr.*, **124**, 411.
535. Schwedt, G. and Bussemas, H. H. (1976), *Chromatographia*, **9**, 17.

536. Schwedt, G. (1976), *J. Chromatogr.*, **118**, 429.
537. Mee, T. J. and Smith, J. A. (1976), in *High Pressure Liquid Chromatography in Clinical Chemistry*, (ed. P. F. Dixon, C. H. Gray, C. K. Lim and M. S. Stoll), Academic Press, London, New York, San Francisco, p. 119.
538. Knox, J. H. and Jurand, J. (1976), *J. Chromatogr.*, **125**, 89.
539. Persson, B.-A. and Karger, B. L. (1974), *J. Chromatogr. Sci.*, **12**, 521.
540. Persson, B.-A. and Lagerstrom, P.-O. (1976), *J. Chromatogr.*, **122**, 305.
541. Imai, K. (1975), *J. Chromatogr.*, **105**, 135.
542. Lagerstrom, P.-O. (1976), *Acta Pharm. Suec.*, **13**, 213.
543. Wahlund, K.-G. and Lund, U. (1976), *J. Chromatogr.*, **122**, 269.
544. Thomas, H., Ponnath, H. and Müller-Enoch, D. (1975), *J. Chromatogr.*, **103**, 197.
545. Yoshida, A., Yoshioka, M., Tanimura, T. and Tamura, Z. (1976), *J. Chromatogr.*, **116**, 240.
546. Chilcote, D. D. and Mrochek, J. E. (1972), *Clin. Chem.*, **18**, 778.
547. Chilcote, D. D. (1972), *Clin. Chem.*, **18**, 1376.
548. Chilcote, D. D. (1974), *Clin. Chem.*, **20**, 421.
549. Graffeo, A. P. and Karger, B. L. (1976), *Clin. Chem.*, **22**, 184.
550. Meek, J. L. and Neckers, L. M. (1975), *Brain Res.*, **91**, 336.
551. Meek, J. L. (1976), *Anal. Chem.*, **48**, 375.
552. Russell, D. H. (1971), *Nature New Biology, (London)*, **233**, 144.
553. Hatano, H., Sumizu, K., Rokushika, S. and Murakami, F. (1970), *Anal. Biochem.*, **35**, 377.
554. Bremer, H. J., Kohne, E. and Endres, W. (1971), *Clin. Chim. Acta*, **32**, 407.
555. Vandekerckhove, P. and Henderickx, H. K. (1973), *J. Chromatogr.*, **82**, 379.
556. Tabor, H., Tabor, C. W. and Irreverre, F. (1973), *Anal. Biochem.*, **55**, 457.
557. Marton, L. J., Russell, D. H. and Levy, C. C. (1973), *Clin. Chem.*, **19**, 923.
558. Gehrke, C. W., Kuo, K. C., Zumwalt, R. W. and Waalkes, T. P. (1974), *J. Chromatogr.*, **89**, 231.
559. Navratil, J. D. and Walton, H. F. (1975), *Anal. Chem.*, **47**, 2443.
560. Abdel-Monem, M. M. and Ohno, K. (1975), *J. Chromatogr.* **107**, 416.
561. Samejima, K. (1974), *J. Chromatogr.*, **96**, 250.
562. Sugiura, T., Hayashi, T., Kawai, S. and Ohno, T. (1975), *J. Chromatogr.*, **110**, 385.
563. Davankov, V. A. and Semechkin, A. V. (1977), *J. Chromatogr.*, **141**, 313.
564. Newton, N. E., Ohno, K. and Abdel-Monem, M. M. (1976), *J. Chromatogr.*, **124**, 277.
565. Young, P. R. and McNair, H. M. (1976), *J. Chromatogr.*, **119**, 569.
566. Schooley, D. A. and Nakanishi, K. (1973) in *Modern Methods of Steroid Analysis*, (ed. E. Heftmann), Academic Press, New York and London, Chapter 2, p 37.
567. Bradford, H. F., Bennett, G. W. and Thomas, A. J. (1973), *J. Neurochemistry*, **21**, 495.
568. Murren, C., Stelling, D. and Felstead, G. (1975), *J. Chromatogr.*, **115**, 236.
569. Unger, K. and Nyamah, D. (1974), *Chromatographia*, **7**, 63.
570. Ersser, R. S. (1976), in *High Pressure Liquid Chromatography in Clinical Chemistry*, (ed. P. F. Dixon, C. H. Gray, C. K. Lim and M. S. Stoll), Academic Press, London, New York, San Francisco, p. 23.

571. Bakay, B. (1975), *Clin. Chem.*, **21**, 1212.
572. Stein, S., Bohlen, P., Stone, J., Dairman, W. and Udenfriend, S. (1973), *Arch. Biochem. Biophys.*, **155**, 203.
573. McHugh, W., Sandman, R. A., Haney, W. G., Sood, S. P. and Wittmer, D. P. (1976), *J. Chromatogr.*, **124**, 376.
574. Edman, P. (1950), *Acta Chem. Scand.*, **4**, 283.
575. Horvath, C. and Melander, W. (1977), *J. Chromatogr. Sci.*, **15**, 393.
576. Schroeder, W. A., Jones, R. T., Cormick, J. and McCalla, K. (1962), *Anal. Chem.*, **34**, 1570.
577. Catravas, G. N. (1964), *Anal. Chem.*, **36**, 1146.
578. Jones, R. T. (1966), *Technicon Symposium*, **1**, 416.
579. Benson, J. V., Jones, R. T., Cormick, J. and Patterson, J. A. (1966), *Anal. Biochem.*, **16**, 91.
580. Glossmann, H., Kottgen, E., Braunitzer, G. and Lackner, B. (1970), *Hoppe-Seyler's Z. Physiol-Chem.* **351**, 409.
581. Frei, R. W., Michel, L. and Santi, W. (1976), *J. Chromatogr.*, **126**, 665.
582. Hansen, J. J., Greibrokk, T., Currie, B. L., Johansson, K. N-G. and Folkers, K. (1977), *J. Chromatogr.*, **135**, 155.
583. Kikta Jr., E. J. and Grushka, E. (1977), *J. Chromatogr.*, **135**, 367.
584. Haller, W. (1965), *Nature* (London), **206**, 693.
585. Hawk, G. L., Cameron, J. A. and Dufault, L. B. (1972), *Prep. Biochem.*, **2**, 193.
586. Crone, H. D., Dawson, R. M. and Morgan Smith, E., (1975), *J. Chromatogr.*, **103**, 71.
587. Regnier, F. E. and Noel, R. (1976), *J. Chromatogr. Sci.*, **14**, 316.
588. Eltekov, Y. A., Kiselev, A. V., Khokhlova, T. D. and Nikitin, Y. S. (1973), *Chromatographia*, **6**, 187.
589. Kudirka, P. J., Busby, M. G., Carey, R. N. and Toren, E. C. (1975), *Clin. Chem.*, **21**, 450.
590. Chang, S-H., Noel, R. and Regnier, F. E. (1976), *Anal. Chem.*, **48**, 1839.
591. Chang, S-H., Gooding, K. M. and Regnier, F. E. (1976), *J. Chromatogr.*, **125**, 103.
592. Schroeder, R. R., Kudirka, P. J. and Toren Jr., E. C. (1977), *J. Chromatogr.*, **134**, 83.
593. Schlabach, T. D., Chang, S.-H., Gooding, K. M. and Regnier, F. E. (1977), *J. Chromatogr.*, **134**, 91.
594. Cohn, W. E. (1955), in *The Nucleic Acids*, (eds. E. Chargaff and J. N. Davidson), Academic Press, New York, Vol. 1 p. 211.
595. Cohn, W. E. (1967), in *Chromatography*, (ed. E. Heftman), Reingold, New York, 2nd Edition, p. 627.
596. Dixon, P. F., Stoll, M. S. and Lim, C. K. (1976), *Ann. Clin. Biochem.*, **13**, 409.
597. Gere, D. R. (1971), in *Modern Practice of Liquid Chromatography*, (ed. J. J. Kirkland), Wiley – Interscience, New York, Chapter 12.
598. Brown, P. R. (1973), *High Pressure Liquid Chromatography, Biochemical and Biomedical Applications*, Academic Press, New York and London.
599. Anderson, N. G., Green, J. G., Barber, M. L. and Ladd, F. C. (1963), *Anal. Biochem.*, **6**, 153.
600. Singhal, R. P. and Cohn, W. E. (1972), *Anal. Biochem.*, **45**, 585.
601. Singhal, R. P. and Cohn, W. E. (1972), *Biochim. Biophys. Acta*, **262**, 565.
602. Anderson, F. S. and Murphy, R. C. (1976), *J. Chromatogr.*, **121**, 251.

603. Hartwick, R. A. and Brown, P. R. (1976), *J. Chromatogr.*, **126**, 679.
604. Horvath, C., Melander, W. and Molnar, I. (1977), *Anal. Chem.*, **49**, 142.
605. Holton, R. A., Spatz, D. M., Van Tamelen, E. E. and Wierenga, W. (1974), *Biochem. Biophys. Res. Commun.*, **58**, 605.
606. Singhal, R. P. (1973), *Biochim. Biophys Acta*, **319**, 11.
607. Kelmers, A. D., Weeren, H. O., Weiss, J. F., Pearson, R. L. Stulberg, M. P. and Novelli, G. D. (1971), *Methods Enzymol.* **20**(c), 9.
608. Roe, B., Marcu, K. and Dudock, B. (1973), *Biochim. Biophys. Acta*, **319**, 25.
609. Khym, J. X. (1974), *Anal. Biochem.*, **58**, 638.
610. Henry, R. A., Schmit, J. A. and Williams, R. C. (1973), *J. Chromatogr. Sci.*, **11**, 358.
611. Burtis, A. C., Munk, M. N. and MacDonald, F. R. (1970), *Clin. Chem.*, **16**, 667.
612. Khym, J. X. (1974), *J. Chromatogr.*, **97**, 277.
613. Khym, J. X. (1975), *Clin. Chem.*, **21**, 1245.
614. Kirkland, J. J., (1970), *J. Chromatogr. Sci.*, **8**, 72.
615. Hartwick, R. A. and Brown, P. R. (1975), *J. Chromatogr.*, **112**, 651.
616. Shmukler, H. W. (1972), *J. Chromatogr. Sci.*, **10**, 137.
617. Charalambous, G. and Bruckner, K. J. (1975), *Tech. Quart. Master Brew. Assoc. Am.*, **12**, 203.
618. Charalambous, G., Bruckner, K. J., Hardwick, W. A. and Weatherby, T. J. (1974), *Tech. Quart. Master Brew. Assoc. Am.*, **11**, 193.
619. Stahl, K-W., Schlimme, E. and Bojanowski, D. (1973), *J. Chromatogr.*, **83**, 395.
620. Brooker, G. (1970), *Anal. Chem.*, **42**, 1108.
621. Pennington, S. N. (1971), *Anal. Chem.*, **43**, 1701.
622. Wildanger, W. (1975), *Chromatographia*, **8**, 42.
623. Breter, H.-J. and Zahn, R. K. (1973), *Anal. Biochem.*, **54**, 346.
624. Pal, B. C., Regan, J. D. and Hamilton, F. D. (1975), *Anal. Biochem.*, **67**, 625.
625. Brown, P. R., Bobick, S. and Hanley, F. L. (1974), *J. Chromatogr.*, **99**, 587.
626. Beardmore, T. D. and Kelley, W. N. (1971), *Clin. Chem.*, **17**, 795.
627. Brown, P. R. (1970), *J. Chromatogr.*, **52**, 257.
628. Scholar, E. M., Brown, P. R., Parks Jr., R. E. and Calabresi, P. (1973), *Blood*, **41**, 927.
629. Rao, G. H. R., White, J. G., Jachimowicz, A. A. and Witkop Jr., C. J., (1974), *J. Lab. Clin. Med.*, **84**, 839.
630. Brown, P. R. and Parks Jr., R. E. (1973), *Anal. Chem.*, **45**, 948.
631. Dean, B. M. and Perrett, D. (1976), *Biochim. Biophys. Acta*, **437**, 1.
632. Brown, P. R. and Miech, R. P. (1972), *Anal. Chem.*, **44**, 1072.
633. Senftleber, F. C., Halline, A. G., Veening, H. and Drayton, D. A. (1976), *Clin. Chem.*, **22**, 1522.
634. Pfadenhauer, K. H. (1973), *J. Chromatogr.*, **81**, 85.
635. Endele, R. and Lettenbauer, G. (1975), *J. Chromatogr.*, **115**, 228.
636. Scott, C. D. (1968), *Clin. Chem.*, **14**, 521.
637. Mrochek, J. E., Butts, W. C., Rainey Jr., W. T. and Burtis, C. A. (1971), *Clin. Chem.*, **17**, 72.
638. Burtis, C. A. (1970), *J. Chromatogr.*, **51**, 183.
639. Gabriel, T. F. and Michalewsky, J. (1972), *J. Chromatogr.*, **67**, 309.
640. Lakings, D. B., Waalkes, T. P. and Mrochek, J. E. (1976), *J. Chromatogr.*, **116**, 83.

641. Breter, H.-J., Seibert, G. and Zahn, R. K. (1976), *J. Chromatogr.*, **118**, 242.
642. Goodman, J. I. (1976), *Anal. Biochem.*, **70**, 203.
643. Evans, N., Games, D. E., Jackson, A. H. and Matlin, S. A. (1975), *J. Chromatogr.*, **115**, 325.
644. Evans, N., Jackson, A. H., Matlin, S. A. and Towill, R. (1976), in *High Pressure Liquid Chromatography in Clinical Chemistry*, (ed. P. F. Dixon, C. H. Gray, C. K. Lim and and M. S. Stoll), Academic Press, London, New York, San Franciso, p. 71.
645. Evans, N., Jackson, A. H., Matlin, S. A. and Towill, R. (1976), *J. Chromatogr.*, **125**, 345.
646. Gray, C. H., Lim, C. K. and Nicholson, D. C. (1976), in *High Pressure Liquid Chromatography in Clinical Chemistry*, (ed. P. F. Dixon, C. H. Gray, C. K. Lim and M. S. Stoll), Academic Press, London, New York, San Francisco, p. 80.
647. Carlson, R. E. and Dolphin, D. (1976), in *High Pressure Liquid Chromatography in Clinical Chemistry*, (ed. P. F. Dixon, C. H. Gray, C. K. Lim and M. S. Stoll), Academic Press, London, New York, San Francisco, p. 87.
648. Battersby, A. R., Buckley, D. G., Hodgson, G. L., Markwell, R. E. and McDonald, E. (1976), in *High Pressure Liquid Chromatography in Clinical Chemistry*, (ed. P. F. Dixon, C. H. Gray, C. K. Lim and M. S. Stoll), Academic Press, London, New York, San Francisco, p. 63.
649. Stoll, M. S., Lim, C. K. and Gray, C. H. (1976) in *High Pressure Liquid Chromatography in Clinical Chemistry*, (ed. P. F. Dixon, C. H. Gray, C. K. Lim and M. S. Stoll), Academic Press, London, New York, San Fransisco, p. 98.
650. Williams, R. C., Schmit, J. A. and Henry, R. A. (1972), *J. Chromatogr. Sci.*, **10**, 494.
651. Vecchi, M., Vesely, I and Oesterhelt, G. (1973), *J. Chromatogr.*, **83**, 447.
652. Puglisi, C. V. and De Silva, J. A. F. (1976), *J. Chromatogr.*, **120**, 457.
653. Stacewicz-Sapuncakis, M., Chang Wang, H.-II. and Gawienowski, A. M. (1975), *Biochim. Biophys. Acta*, **380**, 264.
654. Van de Weerdhof, T., Wiersum, M. L. and Reissenweber, H. (1973), *J. Chromatogr.*, **83**, 455.
655. Stewart, I. and Wheaton, T. A. (1971), *J. Chromatogr.* **55**, 325.
656. Holasova, M. and Blattna, J. (1976), *J. Chromatogr.*, **123**, 225.
657. Jones, G. and DeLuca, H. F. (1975), *J. Lipid Res.*, **16**, 448.
658. Matthews, E. W., Byfield, P. G. H., Colston, K. W., Evans, I. M. A., Galante, L. S. and MacIntyre, I., (1974), *Febs. Lett.*, **48**, 122.
659. Steuerle, H. (1975), *J. Chromatogr.*, **115**, 447.
660. Tartivita, K. A., Sciarello, J. P. and Rudy, B. C. (1976), *J. Pharm. Sci.*, **65**, 1024.
661. Hofsass, H., Grant, A., Alicino, N. J. and Greenbaum, S. B. (1976), *J. Ass. Offic. Anal. Chem.*, **59**, 251.
662. Tomkins, D. F. and Tscherne, R. J. (1974), *Anal. Chem.*, **46**, 1602.
663. Van Niekerk, P. J. (1973), *Anal. Biochem.*, **52**, 533.
664. Vatassery, G. T. and Hagen, D. F. (1977), *Anal. Biochem.*, **79**, 129.
665. Donnahey, P. L. and Hemming, F. W. (1975), *Biochem. Soc. Trans.*, **3**, 775.
666. Williams, R. C., Baker, D. R. and Schmit, J. A. (1973), *J. Chromatogr. Sci.*, **11**, 618.

667. Callmer, K. and Davies, L. (1974), *Chromatographia*, **7**, 644.
668. Wittmer, D. and Haney Jr., W. G. (1974), *J. Pharm. Sci.*, **63**, 588.
669. Williams, A. K. and Cole, P. D. (1975), *J. Agric. Food Chem.*, **23**, 915.
670. Talley, C. P. (1971), *Anal. Chem.*, **43**, 1512.
671. Machulla, H.-J., Laufer, P. and Stoecklin, G. (1974), *Radiochem. Radioanal. Letts.*, **18**, 275.
672. Hatano, H., Yamamoto, Y., Saito, M., Mochida, E. and Watanabe, S. (1973), *J. Chromatogr.*, **83**, 373.
673. Wahlund, K.-G. (1975), *J. Chromatogr.*, **115**, 411.
674. Kirk, J. R. (1977), *J. Assoc. Offic. Anal. Chem.* **60**, 1234.
675. Pachla, L. A. and Kissinger, P. T. (1976), *Anal. Chem.*, **48**, 364.
676. Sood, S. P., Sartori, L. E., Wittmer, D. P. and Haney, W. G. (1976), *Anal. Chem.*, **48**, 796.
677. Lam, K.-W. and Lee, P.-F. (1975), *Invest. Ophthalmol.*, **14**, 947.
678. Eisenbeiss, F. and Sieper, H. (1973), *J. Chromatogr.*, **83**, 439.
679. Zimmerli, B. (1977), *J. Chromatogr.*, **131**, 458.
680. Wheals, B. B., Vaughan, C. G. and Whitehouse, M. J. (1975), *J. Chromatogr.*, **106**, 109.
681. Sparacino, C. M. and Hines, J. W. (1976), *J. Chromatogr. Sci.*, **14**, 549.
682. Laboratory Impex Ltd., Twickenham, Middlesex TW1 4EE. G.B.
683. Laboratorium Prof. Dr. Berthold, D 7547 Wildbad, GFR.
684. Reeve, D. R. and Crozier, A. (1977), *J. Chromatogr.* **137**, 271.
685. Reeve, D. R., Yokota, T., Nash, L. J. and Crozier, A. (1976), *J. Exp. Botany*, **27**, 1243.
686. Ramsteiner, K. A. and Hörmann, W. D. (1975), *J. Chromatogr.*, **104**, 438.
687. Ott, D. E. (1975), *Residue Reviews*, (ed. F. A. Gunther), Springer-Verlag, New York, vol. 55 p. 1.
688. Thornburg, W. (1977), *Anal. Chem.*, **49**, 98R.
689. Horgan, D. F. (1973), *Analytical Methods for Pesticides and Plant Growth Regulators VII, Thin Layer and Liquid Chromatography and Analyses of Pesticides of International Importance*, (ed. J. Sherma and G. Zweig), Academic Press New York, Chapter 2 p. 89.
690. Moye, H. A. (1975), *J. Chromatogr. Sci.*, **13**, 268.
691. Aitzetmüller, K. (1975), *J. Chromatogr.*, **107**, 411.
692. Brinkman, U. A. Th., De Kok, A., Reymer, H. G. M. and De Vries, G. (1976), *J. Chromatogr.*, **129**, 193.
693. Brinkman, U. A. Th. and De Kok, A. (1976), *J. Chromatogr.*, **129**, 451.
694. Brinkman, U. A. Th., De Kok, A., De Vries, G. and Reymer, H. G. M. (1976), *J. Chromatogr.*, **128**, 101.
695. Seiber, J. N. (1974), *J. Chromatogr.*, **94**, 151.
696. Balayannis, P. G. (1974), *J. Chromatogr.*, **90**, 198.
697. Larose, R. H. (1974), *J. Ass. Offic. Anal. Chem.*, **57**, 1046.
698. Zimmerli, B. and Marek, B. (1975), *Mitt. Gebiete Lebensm. Hyg.*, **66**, 362.
699. Little, J. N., Horgan, D. F. and Bombaugh, K. J. (1970), *J. Chromatogr. Sci.*, **8**, 625.
700. Bombaugh, K. J., Levangie, R. F., King, R. N. and Abrahams, L. (1970), *J. Chromatogr. Sci.*, **8**, 657.
701. Majors, R. E. (1973), *Anal. Chem.*, **45**, 759.
702. Beland, F. A., Farwell, S. O. and Geer, R. D. (1974), *J. Agr. Food Chem.*, **22**, 1148.
703. Kirkland, J. J. (1969), *Anal. Chem.*, **41**, 218.

704. Henry, R. A., Schmit, J. A., Dieckman, J. F. and Murphey, F. J. (1971), *Anal. Chem.*, **43**, 1053.
705. Lawrence, J. F., Renault, C. and Frei, R. W. (1976), *J. Chromatogr.*, **121**, 343.
706. Szalontai, G. (1976), *J. Chromatogr.*, **124**, 9.
707. Self, C., McKerrell, E. H. and Webber, T. J. N. (1975), *Proc. Analyt. Div. Chem. Soc.*, **12**, 288.
708. Koen, J. G. and Huber, J. F. K. (1970), *Anal. Chim. Acta*, **51**, 303.
709. Kvalvag, J., Elliot, D. L., Iwata, Y. and Gunther, F. A. (1977), *Bull. Environm. Contam. Toxicol*, **17**, 253.
710. Ott, D. E., (1977), *Bull. Enironm. Contam. Toxicol.* **17**, 261.
711. Jackson, E. R.(1977), *J. Ass. Offic. Anal. Chem.*, **60**, 724.
712. Zehner, J. M. and Simonaitis, R. A. (1977), *J. Ass. Offic. Anal. Chem.*, **60**, 14.
713. Colvin, B. M., Engdahl, B. S. and Hanks, A. R. (1974), *J. Ass. Offic. Anal. Chem.*, **57**, 648.
714. Argauer, R. J. and Warthen Jr., J. D. (1975), *Anal. Chem.*, **47**, 2472.
715. Hosler Jr., C. F. (1974), *Bull. Environm. Contam. Toxicol.*, **12**, 599.
716. Leitch, R. E. (1971), *J. Chromatogr. Sci.*, **9**, 531.
717. Frei, R. W., Lawrence, J. F., Hope, J. and Cassidy, R. M. (1974), *J. Chromatogr. Sci.*, **12**, 40.
718. Frei, R. W. and Lawrence, J. F. (1973), *J. Chromatogr.*, **83**, 321.
719. Corley, C., Miller, R. W. and Hill, K. R. (1974), *J. Ass. Offic. Anal. Chem.*, **57**, 1269.
720. Oehler, D. D. and Holman, G. M. (1975), *J. Agr. Food Chem.*, **23**, 590.
721. Dunham, L. L., Schooley, D. A. and Siddall J. B. (1975), *J. Chromatogr. Sci.*, **13**, 334.
722. Redfern, R. E., Sarmiento, R., Beroza, M. and Mills, G. D. (1975), *J. Econ. Entomology*, **68**, 377.
723. Morgan, E. D. and Poole, C. F. (1976), in *Advances in Insect Physiology. Volume 12.*, (ed. J. E. Treherne, M. J. Berridge and V. B. Wigglesworth), Academic Press, London, p. 17.
724. Smith, A. E. and Lord, K. A. (1975), *J. Chromatogr.*, **107**, 407.
725. Gonnet, C. and Rocca, J. L. (1975), *J. Chromatogr.*, **109**, 297.
726. Lawrence, J. F. (1976), *J. Chromatogr. Sci.*, **14**, 557.
727. Kirkland, J. J. (1969), *J. Chromatogr. Sci.*, **7**, 7.
728. Subach, D. J., Barnes, D. and Wyche, C. (1976), *J. Chromatogr.*, **125**, 435.
729. Kawano, Y., Audino, J. and Edlund, M. (1975), *J. Chromatogr.*, **115**, 289.
730. Still, G. G. and Mansager, E. R. (1975), *Chromatographia*, **8**, 129.
731. Byast, T. H. and Cotterill, E. G. (1975), *J. Chromatogr.*, **104**, 211.
732. Kirkland, J. J., Holt, R. F. and Pease, H. L. (1973), *J. Agr. Food Chem.*, **21**, 368.
733. Kirkland, J. J. (1973), *J. Agr. Food Chem.*, **21**, 171.
734. Maeda, M. and Tsuji, A. (1976), *J. Chromatogr.*, **120**, 449.
735. Wolkoff, A. W., Onuska, F. I., Comba, M. E. and Larose, R. H. (1975), *Anal. Chem.*, **47**, 754.
736. Mundy, D. E., Quick, M. P. and Machin, A. F. (1976), *J. Chromatogr.*, **121**, 335.
737. Fasco, M. J., Piper, L. J. and Kaminsky, L. S. (1977), *J. Chromatogr.*, **131**, 365.

738. Wong, L. T., Solomonraj, G. and Thomas, B. H. (1977), *J. Chromatogr.*, **135**, 149.
739. Rao, G. H. R. and Anders, M. W. (1973), *J. Chromatogr.*, **84**, 402.
740. Seitz, L. M. (1975), *J. Chromatogr.*, **104**, 81.
741. Hsieh, D. P. H., Fitzell, D. L., Miller, J. L. and Seiber, J. N. (1976), *J. Chromatogr.*, **117**, 474.
742. Garner, R. C. (1975), *J. Chromatogr.*, **103**, 186.
743. Pons Jr., W. A. (1976), *J. Ass. Offic. Anal. Chem.*, **59**, 101.
744. Pons Jr., W. A. and Franz Jr., A. O. (1977), *J. Ass. Offic. Anal. Chem.*, **60**, 89.
745. Takahashi, D. M. (1977), *J. Chromatogr.*, **131**, 147.
746. Ware, G. M., Thorpe, C. W. and Pohland, A. E. (1974), *J. Ass. Offic. Anal. Chem.*, **57**, 1111.
747. Rao, G. H. R., Jachimowicz, A. A. and White, J. G. (1974), *J. Chromatogr.*, **96**, 151.
748. Hecht, S. S., Ornaf, R. M. and Hoffmann, D. (1975), *Anal. Chem.*, **47**, 2046.
749. Iwaoka, W. and Tannenbaum, S. R. (1976), *J. Chromatogr.*, **124**, 105.
750. Wolfram, J. H., Feinberg, J. I., Doerr, R. C. and Fiddler, W. (1977), *J. Chromatogr.*, **132**, 37.
751. Singer, M., Singer, S. S. and Schmidt, D. G. (1977), *J. Chromatogr.*, **133**, 59.
752a. *Proceedings of the Second International Symposium on Nitrite in Meat Products. Central Institute for Nutrition and Food Research, TNO, Zeist, the Netherlands, September 7–10, 1976*, (ed. B. J. Tinbergen and B. Krol), Publ. by: Centre for Agricultural Publishing and Documentation, Wageningen, Netherlands, 1977.
752b. *Environmental N-nitroso-compounds; analysis and formation. Proceedings of a Working Conference held at the Polytechnical Institute, Tallinn, Estonian SSR, October 1975*, (eds. E. A. Walker, P. Bogovski and L. Griciute), Publ. by: International Agency for Research on Cancer, Lyon, 1976.
753. Gutmann, H. R., Malejka-Giganti, D. and McIver, R. (1975), *J. Chromatogr.*, **115**, 71.
754. Fiala, E. S., Bobotas, G., Kulakis, C. and Weisburger, J. H. (1976), *J. Chromatogr.*, **117**, 181.
755. Selkirk, J. K., Croy, R. G., Roller, P. P. and Gelboin, H. V. (1974), *Cancer Res.*, **34**, 3474.
756. Selkirk, J. K. Croy, R. G. and Gelboin, H. V. (1975), *Arch. Biochem. Biophys.*, **168**, 322.
757. Holder, G., Yagi, H., Dansette, P., Jerina, D. M., Levin, W., Lu, A. Y. H. and Conney, A. H. (1974), *Proc. Nat. Acad. Sci. USA.* **71**, 4356.
758. Klimisch, H.-J. and Ambrosius, D. (1976), *J. Chromatogr.*, **120**, 299.
759. O'Hara, J. R., Chin, M. S., Dainius, B. and Kilbuck, J. H. (1974), *J. Food Sci.*, **39**, 38.
760. Dong, M., Locke, D. C. and Ferrand, E. (1976), *Anal. Chem.*, **48**, 368.
761. Fox, M. A. and Staley, S. W. (1976), *Anal. Chem.*, **48**, 992.
762. Amos, R. (1972), in *Gas Chromatography, Proc. Internat. Symp.* (Eur), **9**, 235.
763. Suatoni, J. C., Garber, H. R. and Davis, B. E., (1975), *J. Chromatogr. Sci.*, **13**, 367.
764. Suatoni, J. C. and Swab, R. E. (1975), *J. Chromatogr. Sci.*, **13**, 361.
765. Camin, D. L. and Raymond, A. J. (1973), *J. Chromatogr. Sci.*, **11**, 625.

766. Sleight, R. B. (1973), *J. Chromatogr.*, **83**, 31.
767. Vaughan, C. G., Wheals, B. B. and Whitehouse, M. J. (1973), *J. Chromatogr.*, **78**, 203.
768. Haeberer, A. F., Snook, M. E. and Chortyk, O. T. (1975), *Anal. Chim. Acta*, **80**, 303.
769. Karger, B. L., Martin, M., Loheac, J. and Guiochon, G. (1973), *Anal. Chem.*, **45**, 496.
770. Lochmüller, C. H. and Amoss, C. W. (1975), *J. Chromatogr.*, **108**, 85.
771. Popl, M., Dolansky, V. and Mostecky, J. (1976), *J. Chromatogr.*, **117**, 117.
772. Martin, M., Loheac, J. and Guiochon, G. (1972), *Chromatographia*, **5**, 33.
773. Callmer, K., Edholm, L.-E. and Smith, B. E. F., (1977), *J. Chromatogr.*, **136**, 45.
774. Bhatia, K. (1973), *Anal. Chem.*, **45**, 1344.
775. Wolkoff, A. W. and Larose, R. H. (1974), *J. Chromatogr.*, **99**, 731.
776. Wolkoff, A. W. and Larose, R. H. (1975), *J. Chromatogr.*, **111**, 472.
777. Katz, S., Pitt Jr., W. W. (1975), *J. Chromatogr.*, **111**, 470.
778. Katz, S., Pitt Jr., W. W., Mrochek, J. E. and Dinsmore, S. (1974), *J. Chromatogr.*, **101**, 193.
779. Ayres, D. C. and Gopalan, R. (1976), in *High Pressure Liquid Chromatography in Clinical Chemistry*, (ed. P. F. Dixon, C. H. Gray, C. K. Lim and M. S. Stoll), Academic Press, London, New York, San Francisco, p. 195.
780. Chandler, C. D., Gibson, G. R. and Bolleter, W. T. (1974), *J. Chromatogr.*, **100**, 185.
781. Dalton, R. W., Chandler, C. D. and Bolleter, W. T. (1975), *J. Chromatogr. Sci.*, **13**, 40.
782. Doali, J. O. and Juhasz, A. A. (1974), *J. Chromatogr. Sci.*, **12**, 51.
783. Walsh, J. T., Chalk, R. C. and Merritt Jr., C. (1973), *Anal. Chem.*, **45**, 1215.
784. Chandler, C. D. and Bolleter, W. T. (1975), *J. Chromatogr.*, **108**, 365.
785. Farey, M. G. and Wilson, S. E. (1975), *J. Chromatogr.*, **114**, 261.
786. Callmer, K. (1975), *J. Chromatogr.*, **115**, 397.
787. Cox, G. B. (1976), *J. Chromatogr.*, **116**, 244.
788. Wolkoff, A. W. and Larose, R. H. (1975), *Anal. Chem.*, **47**, 1003.
789. Chapman, J. N. and Beard, H. R. (1973), *Anal. Chem.*, **45**, 2268.
790. Dunlap, K. L., Sandridge, R. L. and Keller, J. (1976), *Anal. Chem.*, **48**, 497.
791. Mefford, I., Keller, R. W., Adams, R. N., Sternson, L. A. and Yllo, M. S. (1977), *Anal. Chem.*, **49**, 683.
792. Cassidy, R. M. and Niro, C. M. (1976), *J. Chromatogr.*, **126**, 787.
793. Mori, S. (1976), *J. Chromatogr.*, **129**, 53.
794. Hostettmann, K. and McNair, H. M. (1976), *J. Chromatogr.*, **116**, 201.
795. Rai, P. P., Turner, T. D. and Matlin, S. A. (1975), *J. Chromatogr.*, **110**, 401.
796. Wulf, L. W. and Nagel, C. W. (1976), *J. Chromatogr.*, **116**, 271.
797. Becker, H., Wilking, G. and Hostettmann, K. (1977), *J. Chromatogr.*, **136**, 174.
798. Morot-Gaudry, J. E., Fiala, V., Huet, J. C. and Jolivet, E. (1976), *J. Chromatogr.*, **117**, 279.
799. Rapp, A., Ziegler, A., Bachmann, O. and During, H. (1976), *Chromatographia*, **9**, 44.
800. Sweetser, P. B. and Vatvars, A. (1976), *Anal. Biochem.*, **71**, 68.

801. Carnes, M. G., Brenner, M. L. and Andersen, C. R. (1975), *J. Chromatogr.*, **108**, 95.
802. Challice, J. S. (1975), *Planta*, **122**, 203.
803. Komae, H. and Hayashi, N. (1975), *J. Chromatogr.*, **114**, 285.
804. Ross, M. S. F. (1976), *J. Chromatogr.*, **118**, 273.
805. Glasl, H. (1975), *J. Chromatogr.*, **114**, 215.
806. Kingston, D. G. I., Chen, P. N. and Vercellott, J. R. (1976), *J. Chromatogr.*, **118**, 414.
807. Seitz, L. M. and Mohr, H. E. (1976), *Anal. Biochem.*, **70**, 224.
808. Heyland, S. and Moll, H. (1977), *Mitt. Gebiete Lebensm. Hyg.*, **68**, 72.
809. Schwarzenbach, R. (1977), *Mitt. Gebiete Lebensm. Hyg.*, **68**, 64.
810. Schmit, J. A., Williams, R. C. and Henry, R. A. (1973), *J. Agr. Food Chem.*, **21**, 551.
811. Teitelbaum, C. L. (1977), *J. Agr. Food Chem.*, **25**, 466.
812. Williams, A. T. R. and Slavin, W. (1977), *J. Agr. Food Cnem.*, **25**, 756.
813. Martin, G. E., Guinand, G. G. and Figert, D. M. (1973), *J. Agr. Food Chem.*, **21**, 544.
814. Molyneux, R. J. and Wong, Y. (1973), *J. Agr. Food Chem.*, **21**, 531.
815. Palamand, S. R. and Aldenhoff, J. M. (1973), *J. Agr. Food Chem.*, **21**, 535.
816. Walradt, J. P. and Shu, C-K, (1973), *J. Agr. Food Chem.*, **21**, 547.
817. Hoefler, A. C. and Coggon, P. (1976), *J. Chromatogr.*, **129**, 460.
818. Rapp, A. and Ziegler, A. (1976), *Proceedings of the Königsteiner Chromatographie-Tage*, (Waters GmbH, D-6240, Königstein, BRD), p. 174.
819. Rapp, A. and Ziegler, A. (1976), *Chromatographia*, **9**, 148.
820. Vandercook, C. E. and Price, R. L. (1974), *J. Ass. Offic. Anal. Chem.*, **57**, 124.
821. Palmer, J. K. and List, D. M. (1973), *J. Agr. Food Chem.*, **21**, 903.
822. Schwarzenbach, R. (1976), *J. Chromatogr.*, **129**, 31.
823. Fisher, J. F. (1977), *J. Agr. Food Chem.*, **25**, 682.
824. Nelson, J. J. (1973), *J. Chromatogr. Sci.*, **11**, 28.
825. Wildanger, W. A. (1973), *Chromatographia*, **6**, 381.
826. Battaglia, R. (1977), *Mitt. Gebiete Lebensm. Hyg.*, **68**, 28.
827. Eisenbeiss, F., Weber, M. and Ehlerding, S. (1977), *Chromatographia*, **10**, 262.
828. Singh, M. (1974), *J. Ass. Offic. Anal. Chem.*, **57**, 219.
829. Singh, M. (1974), *J. Ass. Offic. Anal. Chem.*, **57**, 358.
830. Singh, M. (1975), *J. Ass. Offic. Anal. Chem.*, **58**, 48.
831. Bailey, J. A. and Cox, E. A. (1975), *J. Ass. Offic. Anal. Chem.*, **58**, 609.
832. Passarelli, R. J. and Jacobs, E. S. (1975), *J. Chromatogr. Sci.*, **13**, 153.
833. Marmion, D. M. (1977), *J. Ass. Offic. Anal. Chem.*, **60**, 168.
834. Singh, M. (1977), *J. Ass. Offic. Anal. Chem.*, **60**, 173.
835. Jursik, F. (1975) in *Liquid Column Chromatography. A Survey of Modern Techniques and Applications*, (ed. Z. Deyl, K. Macek and J. Janak), Elsevier, Amsterdam, Chapter 51.
836. Kauffman, G. B., Anderson, G. L. and Teter, L. A. (1975), *J. Chromatogr.*, **114**, 465.
837. Kauffman, G. B., Gump, B. H. and Stedjee, B. J. (1976), *J. Chromatogr.*, **118**, 433.
838. Uden, P. C., Henderson, D. E. and Kamalizad, A. (1974), *J. Chromatogr. Sci.*, **12**, 591.

839. Pryde, A. (1978), *J. Chromatogr.*, **152**, 123.
840. Enos, C. T., Geoffroy, G. L. and Risby, T. H. (1977), *J. Chromatogr. Sci.*, **15**, 83.
841. Graf, R. E. and Lillya, C. P. (1973), *J. Organometal. Chem.*, **47**, 413.
842. Greenwood, J. M., Veening, H. and Willeford, B. R. (1972), *J. Organometal. Chem.*, **38**, 345.
843. Eberhardt, R., Glotzmann, C., Lehner, H. and Schloegl, K. (1974), *Tet. Lett.*, 4365.
844. Eberhardt, R., Lehner, H. and Schloegl, K. (1973), *Monatsh. Chem.*, **104**, 1409.
845. Evans, W. J. and Hawthorne, M. F. (1974), *J. Chromatogr.*, **88**, 187.
846. Heizmann, P. and Ballschmiter, K. (1973), *Z. Anal. Chem.*, **266**, 206.
847. Uden, P. C. and Walters, F. H. (1975), *Anal. Chim. Acta*, **79**, 175.
848. Gaetani, E., Laureri, C. F., Mangia, A. and Parolari, S. (1976), *Anal. Chem.*, **48**, 1725.
849. Huber, J. F. K., Kraak, J. C. and Veening, H. (1972), *Anal. Chem.*, **44**, 1554.
850. Uden, P. C., Parees, D. M. and Walters, F. H. (1975), *Anal. Lett.*, **8**, 795.
851. Yoshikawa, Y., Kojima, M., Fujita, M., Iida, M. and Yamatera, H. (1974), *Chem. Lett*, 1163.
852. Enos, C. T., Geoffroy, G. L. and Risby, T. H. (1976), *Anal. Chem.*, **48**, 990.
853. Seymour, M. D. and Fritz, J. S. (1973), *Anal. Chem.*, **45**, 1632.
854. Horwitz, E. P. and Bloomquist, C. A. A. (1974), *J. Chromatogr. Sci.*, **12**, 11.
855. Helmchen, G. and Strubert, W. (1974), *Chromatographia*, **7**, 713.
856. Helmchen, G. (1977), *Proceedings of the Königsteiner Chromatographie Tage* (Waters GmbH, D-6240 Königstein, BRD) p. 85.
857. Furukawa, H. and Sakakibara, F. (1975), *Chem. Pharm. Bull.*, **23**, 1623.
858. Furukawa, H., Mori, Y., Takeuchi, Y. and Ito, K., (1977), *J. Chromatogr.*, **136**, 428.
859. Valentine, D., Chan, K. K., Scott, C. G., Johnson, K. K., Toth, K. and Saucy, G. (1976), *J. Org. Chem.*, **41**, 62.
860. Scott, C. G., Petrin, M. J. and McCorkle, T. (1976), *J. Chromatogr.*, **125**, 157.
861. Sousa, L. R., Hoffman, D. H., Kaplan, L. and Cram, D. J. (1974), *J. Amer. Chem. Soc.*, **96**, 7100.
862. Dotsevi, G., Sogah, Y. and Cram, D. J. (1975), *J. Amer. Chem. Soc.*, **97**, 1259.
863. Dotsevi, G., Sogah, Y. and Cram, D. J. (1976), *J. Amer. Chem. Soc.*, **98**, 3038.
864. Pirkle, W. H. and Sikkenga, D. L. (1976), *J. Chromatogr.*, **123**, 400.
865. Mikes, F., Boshart, G. and Gil-Av, E. (1976), *J. Chromatogr.*, **122**, 205.
866. Grushka, E. and Scott, R. P. W. (1973), *Anal. Chem.*, **45**, 1626.
867. Kikta Jr., E. J. and Grushka, E. (1977), *J. Chromatogr.*, **135**, 367.
868. Hansch, C. (1969), *Acc. Chem. Res.*, **2**, 232.
869. Leo, A., Hansch, C. and Elkins, D. (1971), *Chem. Rev.*, **71**, 525.
870. Federal Register, Part II. Environmental Protection Agency Pesticide Program (25 June, 1975), 40(123) 26880.
871. McCall, J. M. (1975), *J. Med. Chem.*, **18**, 549.
872. Mirrlees, M. S., Moulton, S. J., Murphy, C. T. and Taylor, P. J. (1976), *J. Med. Chem.*, **19**, 615.

873. Haggerty, Jr., W. J. and Murrill, E. A. (1974), *Res. Dev.*, **25**, 30.
874. Henry, D., Block, J. H., Anderson, J. L. and Carlson, G. R. (1976), *J. Med. Chem.*, **19**, 619.
875. Carlson, R. M., Carlson, R. E. and Kopperman, H. L. (1975), *J. Chromatogr.*, **107**, 219.
876. Vitali, T., Gaetani, E., Laureri, C. F. and Branca, C. (1976), Il. Farmaco. Ed. Sc., **31**, 58.

Compound index

Subject index